The Science of Human Origins

The Science of Human Origins

Claudio Tuniz, Giorgio Manzi & David Caramelli

LONDON AND NEW YORK

First published 2014 by Left Coast Press, Inc.

Published 2016 by Routledge
2 Park Square, Milton Park, Abingdon, Oxon OX14 4RN
711 Third Avenue, New York, NY 10017, USA

Routledge is an imprint of the Taylor & Francis Group, an informa business

Library of Congress Cataloging-in-Publication Data:

Tuniz, C. (Claudio)
[Scienza delle nostre origini. English]
Science of human origins / Claudio Tuniz, Giorgio Manzi, and David Caramelli.
pages cm
Summary: "Our understanding of human origins has been revolutionized by new discoveries in the past two decades. In this book, three leading paleoanthropologists and physical scientists illuminate, in friendly, accessible language, the amazing findings behind the latest theories. They describe new scientific and technical tools for dating, DNA analysis, remote survey, and paleoenvironmental assessment that enabled recent breakthroughs in research. They also explain the early development of the modern human cortex, the evolution of symbolic language and complex tools, and our strange cousins from Flores and Denisova."—Provided by publisher.
Includes bibliographical references and index.
ISBN 978-1-61132-971-1 (hardback)—ISBN 978-1-61132-972-8 (paperback)—ISBN 978-61132-973-5 (institutional eBook)
1. Paleoanthropology. 2. Human evolution. 3. Fossil hominids. I. Manzi, Giorgio. II. Caramelli, David. III. Title.
GN281.T8513 2013
569.9—DC23

2013041025

Published in Italian as *La scienza delle nostre origini,* Rome: Editori Laterza, ISBN 9788858106716
Translation revised by Peter Mcgrath, The World Academy of Sciences, Trieste, Italy

ISBN 978-1-61132-971-1 hardcover
ISBN 978-1-61132-972-8 paperback

Cover design by Jane Burton
Cover image: CT scan of Neanderthal skull by Fabio Di Vincenzo (© Giuseppe Sergi Anthropology Museum)

Contents

Illustrations

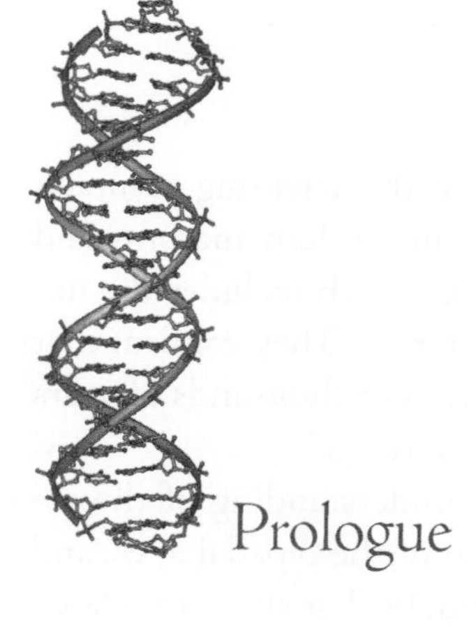

Prologue

Physical sciences have recently intensified their incursions into the study of human origins, thus going far beyond their traditional focus on the origins of matter and of the universe. Human evolution research fuels heated debates with philosophers and theologians, as it raises a number of questions at the margins of science. What does it mean to be human? What are the cognitive boundaries between us and other now extinct human species? What are the philosophical and ethical implications of the former coexistence of different human species? Are religious and ethical attitudes the product of cultural or rather biological evolution?

This book will focus on the new scientific evidence available to reconstruct the story of our evolution. Thanks to physics but also to genetics and other disciplines, paleoanthropology—the science of human origins—has moved from a descriptive to a quantitative approach. As an introduction to the discussion to follow, chapter 1 provides a short review of what paleoanthropologists have discovered in the last 150 years, highlighting some critical issues and points of disagreement. Chapter 2 presents the advanced nuclear chronometers that can measure the pace of human evolution with increasing accuracy. They help us determine when we acquired our modern anatomy and when we developed a fully modern behavior, which includes the use of complex language and symbolic representation. They confirm that several species of *Homo* have coexisted for tens of thousands of years in the past, with possible different "behavioral traits."

There are, of course, huge gaps in our understanding of the selective mechanisms that led to the evolution of the bipedal apes, and later of the human genus; but there is no doubt that they are related to the environment in which these species lived, ate, and multiplied. Unfortunately data on climate change over time and space are not sufficiently robust to test hypotheses relating human evolution to exogenous factors. In chapter 3 we discuss new research programs on past environmental conditions based on the analysis of isotopic abundances and other climate indicators in sediments, stalactites, and Antarctic ice sheets. Such programs should be complemented by the identification of new sites with ancient human remains. To this aim, a systematic use of remote sensing techniques, through satellites, space shuttles, airborne scanners, and ground-penetrating radars, could expose unexplored areas of interest.

Chapter 4 introduces "virtual paleoanthropology": the noninvasive X-ray 3D imaging of fossil remains. This method allows scientists, for example, to reconstruct the brain structure of archaic humans and proto-humans with great precision. The European Synchrotron Radiation Facility in Grenoble recently used this technique to analyze a new species of South African *Australopithecus*, showing that the reorganization of the brain started with this bipedal creature, even though the volume of its brain was more similar to that of an ape. Scientists can thus infer new details on the evolution that led to the development of the brain (and subsequently the mind) in modern humans.

Chapter 5 illustrates the use of population genetics in the study of human origins and dispersal. It also provides basic information on the new science of "paleogenetics," the DNA analysis of ancient fossil remains based on advanced sequencing technology. Chapter 6 discusses the application of these methods to decode the nuclear genome of Neanderthals and of previously unknown humans recently discovered in Siberia. This research aims to identify genes that express our biological nature, including our cognitive and behavioral characteristics.

Finally, a list of notes and a glossary will help the reader with more details on the scientific rationale, methods, and techniques discussed in the book.

The story of our evolution is riddled with mysteries and controversies that excite researchers and the public. However, it is our opinion that only the best paleoanthropological and archaeological science will provide us with reliable answers. Besides, physicists, chemists, and morphologists will contribute with their new chronometers, microscopes, and algorithms; geneticists will provide their high-tech sequencing machines; geologists will be exploring and reconstructing past environments; and ecologists will be modeling the dynamics of ancient human and pre-human populations. This book attempts to provide a brief account of our understanding of human origins on the basis of exciting new scientific methods that can finally help us unravel the greatest puzzle of humankind.

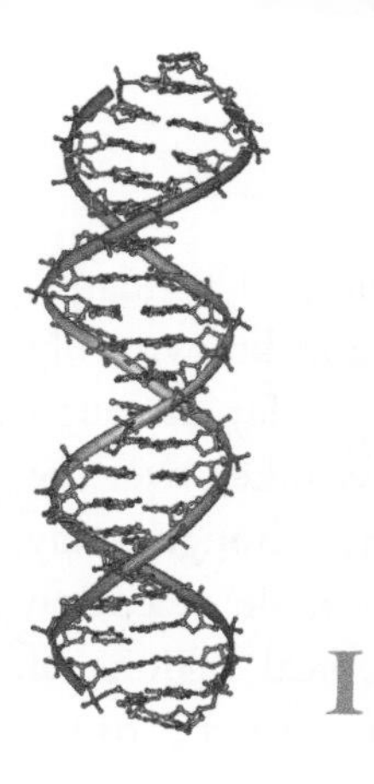

1

Human Origins: Just a Primer

We, the Primates

The origin of the species *Homo sapiens*[1] (our own species, as it was classified by Linnaeus in 1758) is just another episode, in some ways a bit bizarre, in the natural history of a group of creatures known (after Linnaeus) as the primates. This group is an order of the class Mammalia, which includes lemurs, monkeys, and apes. As such, we, the primates, bear imprinted in our physical appearance, in our physiology, and in our DNA all the characteristics that are shared by these kinds of vertebrates. Indeed, we can regulate our body temperature and usually our bodies are covered with (more or less visible) hair; specifically, we have infants that develop inside the mother's body and are nurtured for a certain period after birth.

The fossil record tells us that the story we humans have in common with the other primates dates back to earlier than 65 MYA,[2] at the end of the Mesozoic, the age of the dinosaurs. Since then, small nocturnal animals—insect eaters, with pointed noses and hairy bodies—emerged from the common stock of all primitive mammals, giving

---

rise to the adaptive radiation that led to a huge variety of primates. Although many of these primates are now extinct and known only from their fossils, more than 400 species are still living today. Through the colonization of forest environments, primates have taken over the entire planet and developed a myriad of forms. In fact, although today primates are distributed almost entirely (the most notable exception being our own species, *H. sapiens*) within the tropical regions of South America, Africa, and Southeast Asia, fossils of extinct prosimians, monkeys, and apes have been found also in North America, Europe, and northern Asia.

An examination of the evolutionary history of the primates is beyond the scope of this book. But we must remember that after 35 MYA there was a remarkable expansion of the group of primates known as the Hominoidea: the apes. The current representatives of this group are a few taxa that evolved from a number of other extinct forms and that exhibit many affinities (both morphological and genetic) with our own species: the gibbons (thirteen species), the Asian great apes (the orangutans), and the African apes (gorillas and chimpanzees). The origin of extant Hominoidea, including ourselves, is documented by a large amount of fossil evidence scattered throughout Africa and Eurasia during the Miocene. It is in this context that we must look for taxa close to the origin of our direct ancestry.

From this point of view, one of the hypotheses formulated in 1871 by Charles Darwin[3] proved to be valid. Darwin believed that the origins of *H. sapiens* were to be found in Africa, since the African great apes (gorillas and chimpanzees) are more similar to us than the Asian orangutans and all other primates. In the last decades, molecular biology has fully confirmed this intuition, adding a quantitative approach (unimaginable in Darwin's time) to assess the evolutionary divergence dates with a method known as "molecular clock," as we will discuss in other sections of the book. Thus we know that the common ancestor between the chimpanzees and us lived around 6 MYA; the separation from the gorillas occurred a little earlier (about 8 MYA), and the orangutans separated from this lineage before 14 MYA. This implies that at some point in time some kind of African proto-ape has divided into different populations isolated from one another, probably under the influence of environmental changes. Some of these groups evolved into different species, giving rise to the trajectories of the extant gorillas, chimpanzees, and human beings. Nevertheless, the late Miocene ape from which this story begins has yet to be identified on the basis of the fossil record. What we know for sure is that among its descendants, the ape that is the most distant from the original forest and arboreal adaptation is *H. sapiens,* the only surviving species of a lineage that began to develop around 6 MYA.

Still in the Forest, but . . .

Based on the fossil record of the last 6 or 7 million years, we can identify about twenty extinct species of humans and proto-humans, each representing more or less direct ancestors of today's humankind. The oldest ones were distributed throughout southeastern Africa, from the Horn of Africa down to the Cape. They were found mostly in Ethiopia, Kenya, Tanzania, and South Africa, with two exceptions, both located around 2,000 kilometers to the northwest in Chad.

So, for a few million years (between 6 and 2 MYA), our evolutionary history took place mainly in the geographic area marked by the Great Rift Valley (GRV), the well-known tectonic trench characterizing East Africa, as well as in South Africa and (perhaps more occasionally) in Chad. The development of the GRV has induced the rise of mountains and the creation of deep depressions, which are punctuated by large lake basins. This area has been selectively affected by the progressive drying out that has characterized the history of our planet over the last 10 million years, especially during the last six. Today at the same latitude, but in West Africa, there are dense equatorial forests where chimpanzees and gorillas survive. To the east there are now large extensions of grassland (savannah), dotted with trees and shrubs or interspersed with strips of forest; it is here that our lineage took its first steps.

It must be remarked that the drying trend in the territories east of the GRV was gradual but with distinct periods of accelerated drying. This resulted first in the fragmentation of the existing forests (after 6 MYA) and later in the formation of extended grasslands (around 3 MYA or a little later). The repercussions of these environmental changes are clearly visible in the history of human evolution in terms of both habitat fragmentation (favorable to the emergence of new species) and selective pressure. In fact, it was just around 6 MYA that the human lineage separated from that of the chimpanzees, but it was only after 3 MYA that the essential divergence that led to the emergence of the genus *Homo* occurred.

Over the past couple of decades, the remains of possible earliest ancestors of the human lineage have appeared on the scene. Taken together, they cover the original time span of our history, ranging from less than 7 MYA to more than 4 MYA. They include three genera (and related species) of still hypothetical bipedal apes and ancestors of ours: *Sahelanthropus*, *Ardipithecus*, and *Orrorin*. Let's start with the most ancient of them. In 2002 *Nature* published the preliminary description, taxonomic attribution, and phylogenetic interpretation of a skull found in Chad. This fossil specimen was attributed to a genus and a species hitherto unknown: *Sahelanthropus tchadensis* (nicknamed Toumaï by its discoverers). The skull, almost complete, and a few other remains were found in the area of Lake Chad, now a desert thousands of miles west of the GRV. The chronology, not very precise (deduced from the faunal assemblage only), was based on a stratigraphic horizon ranging between 7 and 6.5 MYA in terms of biochronology. Since then, the morphology of this skull has been the object of numerous studies, but not all paleoanthropologists believe this is an ancestral form of the human lineage. If its age were confirmed, it would exceed the limit set by the molecular clock, which corresponds to about 6 MYA. We could then assume that the calibration based on genetic data is not precise and the separation of our evolutionary line from chimpanzees should be moved further back.

Two years earlier another new species (and new genus), *Orrorin tugenensis*, had been suggested as a possible ancestor of the human lineage, although this hypothesis was based on a few fragmentary fossil remains found in deposits of the Tugen Hills (Kenya) dated to about 6 MYA. If the cranial and dental remains of *Sahelhanthropus* did not allow a reliable diagnosis on its bipedalism (the "trademark" of our evolutionary line), those of *Orrorin*—a couple of well-preserved portions of femur, in particular—suggested a rather advanced bipedalism. Only the discovery of further remains and/or new analyses of those already discovered may shed light on this leading character in the early history of our origins.

Let's finally introduce the genus *Ardipithecus*, discovered in Ethiopia in the area known as Middle Awash, two species of which have been known to us for more than ten years. The oldest one, *Ar. kadabba*, is represented by a very small sample of fragmentary remains with ages ranging between 5.8 and 5.2 MYA; in this case, too, we should suspend our judgment, given the incompleteness of the findings. On the other hand, another species—reliably dated to 4.4 MYA and named (in 1994) *Ar. ramidus*—has been identified through several fossil remains, including many bones from a single skeleton. It is the oldest skeleton among those (actually very few) available to paleoanthropologists, and the media nicknamed it "Ardi." Thanks to advanced analyses, based also on X-ray imaging, the different bones were described in detail in several papers that made up an entire issue of the international journal *Science* (October 2, 2009). What emerged was a mixed picture in which, notwithstanding the persistence of ape-like features—the skull, body proportions, and many anatomical details (such as the opposable big toe)—the evidence suggested placing this species at the roots of our evolutionary tree. Such evidence included some dental traits and, even more importantly, bones of the pelvic girdle that are characteristic of a bipedal ape.

New research will add more fossils, reliable data, and better interpretations to the evidence gathered in recent years. Anything related to the first bipedal apes is of paramount interest to both the specialists and the public, as it concerns the origin of an evolutionary trajectory that millions of years later would lead to the emergence of our own species.

Australopithecus & Friends

We know *Australopithecus* much better than the previous hominids. Remains of at least four species of this extinct genus have been found. Not counting some uncertain (*A. bahrelghazali*) and controversial (*A. garhi*) cases, the most reliable species at the moment are, in chronological order, *A. anamensis, A. afarensis, A. africanus*, and the most recent, *A. sediba*.

A. anamensis is the most ancient *Australopithecus* specimen, since the fossils attributed to this species—from both Kenya (Lake Turkana) and Ethiopia (Middle Awash)—are dated between 4.2 and 3.9 MYA. Its remains well document both bipedal locomotion and other characteristics, mostly dental, that are typical of the genus *Australopithecus*. These and other traits are more clearly expressed in *A. afarensis*, the species that includes the famous skeleton known as "Lucy" (AL 288-1). In fact, this is the best-known variety of *Australopithecus*, thanks to a considerable number of fossil finds distributed across various sites in Ethiopia (e.g., Hadar) and Tanzania (Laetoli) that span a long time period, between approximately 4.0 and 3.0 MYA. The footprints of three bipedal creatures, older than 3.5 MYA, that remained impressed in a layer of volcanic ash found at Laetoli are also credited to *A. afarensis*.

The number and variety of fossil remains attributed to *A. afarensis* allows us to address issues related to the variability of the species and to clarify some aspects of biology and behavior that are extendable to the so-called australopithecines as a whole. These were bipedal apes of medium-large size (just over 5 feet tall, weighing around 35–40 kilograms) that can be easily distinguished from present-day apes, although they had a relatively small cranium (their brain size did not exceed, on average, half a liter) and a rather large face projecting forward, with developed jaws. They did not have the dental proportions of other Hominoidea (that have large protruding incisors and canines, but relatively small premolars and molars); instead, the species of the genus *Australopithecus*—and even more so those of the genus *Paranthropus*, which we will meet soon—had relatively small front teeth, non-protruding canines, and large back teeth.

Another important aspect concerns the upright bipedal posture, a model of locomotion very rare, if not absent, among extant and extinct primates. The late Miocene apes experienced forms of bipedalism, and also present-day chimpanzee can do this for short distances; but the human lineage, including australopithecines, has a gait and body shape that are unique.[4] Bipedalism has affected the entire body, changing nearly any part of the skeleton. The pelvic girdle was particularly affected, with crucial consequences on several aspects related to birth and in evo-devo relationship with brain development, which will appear with the evolution of the genus *Homo* (as we shall see in chapter 4). Furthermore, the upright posture freed the upper limbs from their locomotion commitments, and hence the hands of the previously arboreal primate became capable of manipulating objects with particular dexterity and subsequently producing archaeologically visible artifacts.

In addition to the australopithecines from eastern Africa (*A. anamensis* and *A. afarensis*) and other similar taxa that deserve further attention,[5] some species of *Australopithecus* have been discovered in the southern part of the continent. *A. africanus* has been found in the

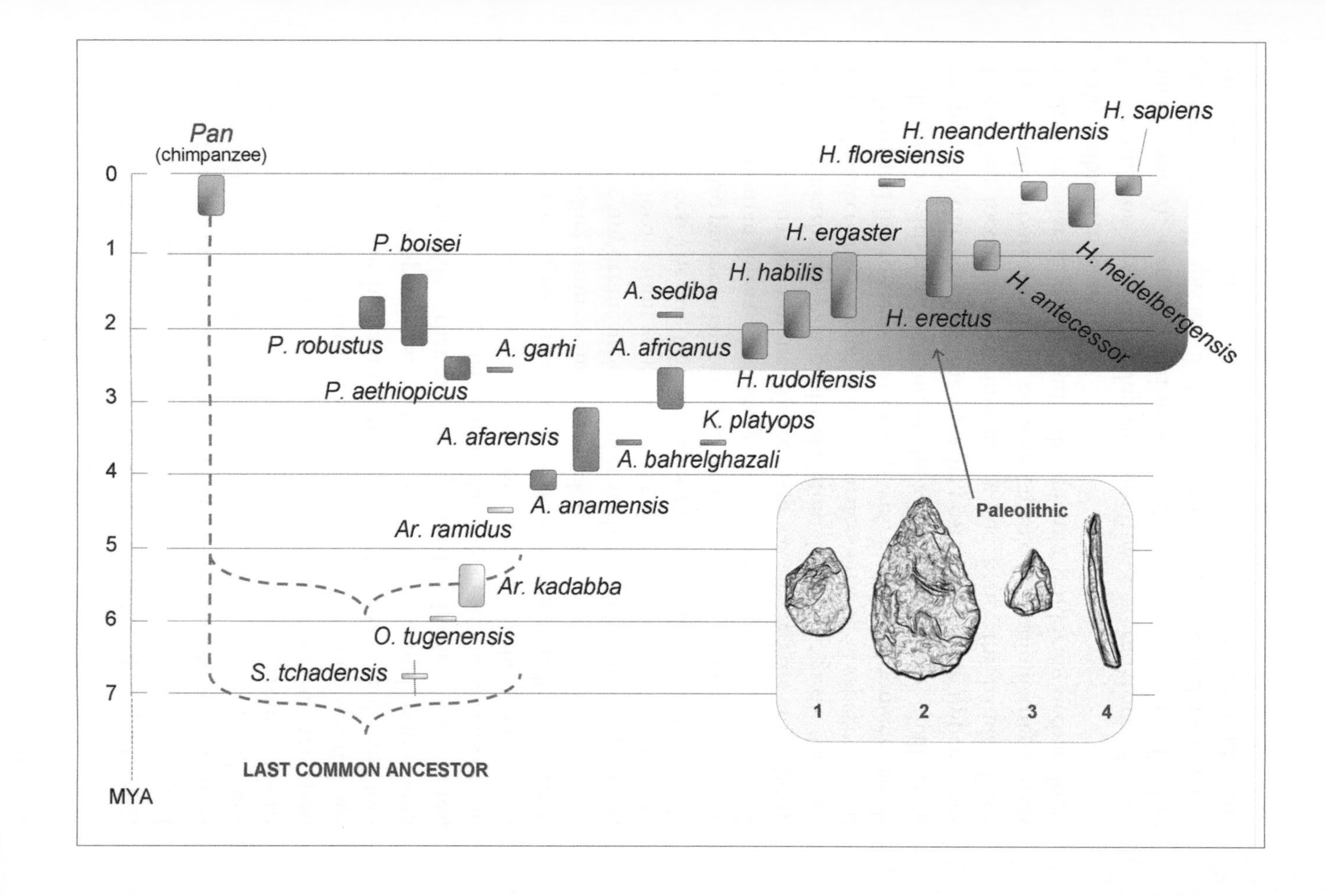

Pan
(chimpanzee)
0
1
2
3
4
5
6
7
MYA
P. boisei
P. robustus
P. aethiopicus
A. garhi
A. sediba
A. africanus
H. habilis
H. ergaster
H. floresiensis
H. neanderthalensis
H. sapiens
H. erectus
H. antecessor
H. heidelbergensis
H. rudolfensis
K. platyops
A. afarensis
A. bahrelghazali
A. anamensis
Ar. ramidus
Ar. kadabba
O. tugenensis
S. tchadensis
LAST COMMON ANCESTOR
Paleolithic
1
2
3
4

sites of Taung, Sterkfontein, and Makapansgat in South Africa. Owing to the characteristics of these deposits (karst fills), it is not easy to attribute precise dates to the remains of this species, but we know they cover a period between about 3 and 2.5 MYA. In this case, too, the fossil remains are of several individuals, and thus we know quite well the characteristics of *A. africanus*. Nevertheless, its exact phylogenetic position remains to be confirmed. Some interpret this species as the last *Australopithecus* form preceding the genus *Homo*; others believe its geographical location disproves this hypothesis, since the oldest evidence of *Homo* comes from eastern (and not southern) Africa, as we shall see shortly.

This debate has been recently reinvigorated by the discovery of the remains of at least four skeletons (over 200 fossils so far) attributed to another australopithecine, *A. sediba*, a species known to the public only since 2010. The numerous and well-preserved remains come from the site of Malapa, 45 kilometers from Johannesburg, South Africa, and are well dated to about 1.9 MYA. The combination of some characteristics common in *Australopithecus* (e.g., the size of the molars) and others closer to the genus *Homo* (e.g., the conformation of the pelvic bones) makes these findings a discovery of enormous interest. In fact, *A. sediba* might play a special role in the origins of our own genus. This is obviously a crucial step in human evolution, for which we have direct (fossil remains) and indirect (artifacts) evidence showing the occurrence, hundreds of thousands of years earlier, of early *Homo* in eastern Africa.

FIGURE 1.1

(*Opposite*) Extinct *hominin* species described on the basis of the fossil record. These species connect *H. sapiens* to extant apes through a bushy path lasting 5–6 million years. The figure includes also a schematic representation of the Paleolithic "modes." LEGEND: S.= Sahelanthropus, O.= Orrorin, Ar.= Ardipithecus, A.= Australopithecus, K.= Kenyanthropus, P.= Paranthropus, H.= Homo. Redrawn from G. Manzi, *L'evoluzione umana* (Bologna, It.: il Mulino 2007).

Paranthropus and Early *Homo*

We already mentioned that, after around 3 MYA, a much drier climate hit the regions near the GRV. This led to a further extension of the savannah in this area, similar to what occurred north and south of the equatorial zone. Our ancestors were able to venture into these open plains stripped of forests, probably taking advantage of the evolutionary adaptations they had previously acquired.[6] The limited food resources available in the new ecosystems triggered additional selective pressures. The fossil evidence suggests that individuals and populations were selected according to their capacity to adopt two very different survival strategies: the first concerns the evolution of the genus *Paranthropus*, and the second concerns that of the genus *Homo*.

We know three species of *Paranthropus*, which lived between about 2.5 and 1.2 MYA. They take the name of *P. aethiopicus*, *P. boisei*, and *P. robustus*. The first two lived in eastern Africa—including Ethiopia, Tanzania, and the region of Lake Malawi—in successive periods between 2.6 and 2.3 MYA and between 2.3 and 1.2 MYA, respectively. By contrast, *P. robustus* lived exclusively in southern Africa in more recent times, between 1.9 and 1.6 MYA. The morphology of the most ancient species, *P. aethiopicus*, is (not surprisingly) the most archaic, probably representing the ancestral form of the other two species.

The three species show the same basic morphology of other australopithecines, but more specialized, especially as far as the masticatory apparatus is concerned: large posterior teeth, big jaws and cheek bones, and a sagittal crest on top of the skull for the insertion of large chewing muscles. For these reasons, they have been termed "robust australopithecines." The teeth exhibited an extreme variation of the model already seen in *Australopithecus*, with reduced front teeth (incisors and canines) and very large premolars and molars. At the same time, the impressive cranial structures suggest a great masticatory efficiency. It seems obvious to conclude that we are dealing with an adaptation to particular feeding strategies influenced by the scarcity of available resources in the new environment of the savannah. Given the scarcity of leaves and fruits, the species of *Paranthropus* adopted diets based on tough, low-quality food such as seeds and roots, which were difficult to chew but were more readily available.

There was, however, another type of food that could be a useful resource to survive in the arid savannah environment: meat. The evidence shows that the first representatives of the genus *Homo* adopted this crucial strategy of survival. At first they become scavengers, exploiting the carcasses made available by other large predators. Hunting activities will arrive later, followed by the use of improved stone tools and fire.

We have known the first representatives of the genus *Homo* since 1964, when the species *H. habilis* was defined on the basis of a number of fragmentary remains found in the Olduvai Gorge, in Tanzania, with an age of about 2 million years. The classification was based on some morphological characteristics, including braincase dimensions (which exceed a volume of half a liter) and smaller posterior teeth, showing a sort of reversal of the trend observed in australopithecines. But there are behavioral aspects that need to be taken into account in addition to the phenotype. In fact, the oldest species of the genus *Homo* was classified using the term "*habilis*," because it was found in the same geological strata associated with the oldest stone artifacts. This is the beginning of the Paleolithic and, more specifically, the "Mode 1" of the Lower Paleolithic (also called Oldowan). Since then, finds of fossil remains and artifacts related to early *Homo* have increased in number and quality, adding further complexity to our understanding of the nature and phylogenetic position of these first, primordial humans. Although there are no serious doubts that they represent the origin of the human lineage, some paleoanthropologists wonder whether they should really be considered the beginning of the genus *Homo* or should still be referred to the genus *Australopithecus*.

The first two *Homo* species are known as *H. habilis* and *H. rudolfensis*. They first populated southern Ethiopia, Kenya, and Tanzania, but it is possible that soon one or both species moved farther south (down to South Africa) as well as north and out of the continent. The dating of Oldowan artifacts found so far in Africa goes from 2.6 to around 1.5 MYA. Morphologically, strong contrasts emerge by comparing the fossil skeletal remains of these first *Homo* species. *H. habilis* had a modest cranial capacity (about 600 milliliters, on average) and teeth of reduced size (typical of *Homo*), but its body proportions hint at a locomotion model that was partly arboreal. By contrast, *H. rudolfensis* had teeth and facial features similar to *Australopithecus*, but a larger brain size (up to about 700 milliliters) and elements of the skeleton more similar to subsequent forms of the genus *Homo*.

The Many Faces of the Genus *Homo*

From the previous discussion, it is clear that our phylogenetic tree has a relatively simple structure. Apart from a number of secondary branches, it resembles a Y, with a main bifurcation where the genus *Paranthropus* and the genus *Homo* started to diverge around 2.5 MYA. Thus, one of the two main branches is the result of the emergence and further differentiation of genus *Homo*. However, the course of our evolution does not proceed linearly across the Pleistocene but becomes more complex, especially because it develops not only in Africa but also on other continents.

In fact, populations of early *Homo* spread immediately (in geological terms) through Eurasia; this led to the evolution of a variety of different human morphs in different regions—some in chronological sequence, but most of them coexisting during the same periods and even within the same regions. These different human populations would meet at times, giving rise to possible competition for resources or, alternatively, to interbreeding. Therefore, the rest of our evolutionary story is far from being gradual and sequential, and it is incompatible with the twentieth-century paradigm about human evolution that is usually referred to as the "single-species hypothesis." According to this point of view, until a recent past the evolution of the genus

Homo was interpreted as a succession of forms. Starting from a unique archaic species diffused through Africa and into Eurasia (*H. erectus*), the largely dispersed human varieties would have eventually produced the present diversity of *H. sapiens* across the planet.

However, this is not what really happened, and nowadays many paleoanthropologists are inclined to describe a number of species of the genus *Homo* (about ten, at present) that overlapped, diverged, and sometimes merged again within a complex scenario. Nonetheless, it is possible that often we are dealing with geographical and/or chronological varieties (or "subspecies") that were not distinct (at least, not entirely) from one another reproductively.[7]

The first of these species (or varieties) has, once again, an African homeland: indeed, it inhabited an area that is often called the "cradle of humankind"—between the GRV and South Africa—for presumably a very long period, between more than 1.8 and less than 1 MYA. It is sufficiently distinct from other similar forms of the genus *Homo*[8] and has been named *H. ergaster*. The specific designation *ergaster*, which in Latin means "craftsman," refers (as *habilis*) to the ability of these archaic humans to build tools. It is this species that clearly marks the advent of a new stage of the Lower Paleolithic— "Mode 2" (also known as Acheulean)—that appears in the African archaeological record around 1.4 MYA. These Paleolithic assemblages are characterized by more refined techniques of stone-knapping and by the recurrent presence of an almond-shaped tool of various dimensions made from different kinds of rocks and also from bone: the so-called handaxe or biface.

Analyzing the morphology of fossil crania attributed to *H. ergaster*, we notice the emergence of a new structural design that will characterize all the species and varieties of the genus *Homo*, with the exception of *H. sapiens*: the cranial vault is low and elongated anteroposteriorly, massive enough to contain (in *H. ergaster*) a brain of about 800 milliliters, which will grow exponentially and reach and exceed a volume of 1500 milliliters in more recent species with large brains, such as the

Neanderthals. This species is also characterized by a number of bone superstructures, especially in the temporal and occipital region, and by a sort of brim above the orbits: the "supraorbital torus." It may be surprising that some elements of "modernity" are already present in the trunk and limbs (as clearly shown by the skeleton known as "Turkana Boy"). For example, adult *H. ergaster* could reach a height of 180 centimeters for an estimated weight of about 70 kilograms. In addition, body proportions and the shape of the pelvis are compatible with bipedalism, which had become the only form of locomotion.

As we know, almost immediately after its evolution in Africa, the genus *Homo* became the protagonist of an unprecedented geographical dispersal, so as to reach rapidly—probably less than 1.5 MYA—the most eastern edges of Asia. In particular, on the island of Java, which was connected to the rest of Indonesia during periods of marine regressions (glaciations), fossil remains have been found that have traits similar to those of *H. ergaster* but that appear to be sufficiently distinct from their African "cousins" to deserve a different name: *H. erectus*.

In Java we have evidence that *H. erectus* lasted possibly between 1.5 MYA and 150 KYA, but both limits need to be verified). During this long lapse of time, the morphology of the fossils remained relatively stable, accompanied by forms of endemism and the progression to a moderate expansion of the endocranial volume: from about 900 milliliters in the oldest crania (for example, those from Sangiran) to 1,200 milliliters in the most recent ones (from Ngandong). Fossil remains from China—found in the sites of Lantian, Zhoukoudian, Hexian, and others—are also largely attributed to *H. erectus*. The most important Chinese site is the cave of Zhoukoudian, about 40 kilometers from Beijing. This site was frequented by *H. erectus* for a long time, with a high concentration of fossil and archaeological evidence around 400 KYA (Mode 1). But how did they disperse from Africa to East Asia?

In the last twenty years, discoveries of fossils in various Asian localities have suggested, even if the dates are sometimes uncertain and some findings are not always convincing, that the genus *Homo* had a tendency to disperse that our prehuman ancestors did not have. Many factors may have contributed to this change: the constant transformation of the environment, which had an impact on populations no longer tied exclusively to a forest habitat; the new "ecological niche" of the early specimens of *Homo* in their new role of scavengers; the changes in their biology and behavior; their successful adaptation and, therefore, demographic expansion; the potential "cultural" development induced by the ability to produce artifacts; and the capability to adapt the natural environment to their needs (yet to be established for such a distant past) with the use of fire, the frequentation of caves, the use of animal skins, etc. At the beginning the dispersal of *Homo* followed a "southern route," remaining initially in the proximity of the tropics, probably because of biological constraints (such as skin pigmentation).

The most remarkable site, one of the few that document the earliest phase of this first dispersal of *Homo*[9] out of Africa, is located on the southern slopes of the Caucasus, in Georgia, and takes its name from the nearby village of Dmanisi. It is a paleontological deposit of great interest, which has revealed skulls, jaws, and postcranial bones of a very primitive *Homo*, associated with Paleolithic artifacts of Mode 1 as well as numerous fossil remains of a fauna reminiscent of that of some African regions. The chronology is equally surprising: these remains date back to more than 1.7 MYA. On the basis of this and other evidence, it seems increasingly likely that the earliest dispersal from Africa involved extremely archaic forms of the genus *Homo* (something between *H. habilis* and *H. ergaster*, with some features of *H. erectus*). The morphological diversity observed among the human fossil samples from Dmanisi (including the present five crania, mandibles, and several postcranial remains) is in fact considerable: it might correspond to a period of great morphological instability at the very beginning of the genus *Homo*. In a sense, this is the fossil sample that would give rise to the evolution of successive humans in Eurasia, including the species *H. erectus* documented in Java and China; possibly

H. antecessor, known so far only in northern Spain; but perhaps also the diminutive humans found on the island of Flores, in Indonesia (the archipelago of Sunda).

The humans found on Flores represent another species of the genus *Homo* (*H. floresiensis*), whose remains have been known to the public only since 2004, and they were nicknamed "hobbits" because of the small brain combined with short stature and disproportionately long arms and feet. Due to its pelvic shape, the reference skeleton, LB1, found in the Liang Bua cave on the island of Flores, was attributed to a female. Many characteristics, especially in the skull, appear similar to those of the early *Homo* fossils found in Dmanisi. Based on their morphology—extremely primitive in many features of the skull and the postcranial skeleton and in their tiny size (height did not exceed one meter, with a brain volume of approximately 400 milliliters)—they might be the legacy of an occasional occupation of the island, which lies halfway between Indonesia and Australia, at the time of the first dispersal of *Homo* into Asia. A long time of isolation and the island ecology would then drive their evolution in a very different direction from the one followed by other *Homo* populations distributed on large continental masses. This phenomenon is a known as "insular dwarfism." There is also a minority of scholars (the so-called skeptics) who insist on the thesis that these "hobbits"[10] are nothing but the representatives of a pigmy population of *H. sapiens*, with one of them affected by microcephaly.

FIGURE 1.2
(*Opposite*) Evolution of the genus *Homo* over a geographical context including the main continental areas. Three major out-of-Africa dispersals can be considered: the first starting about 2 MYA (subsequent to the emergence of the genus *Homo*), the second after 1 MYA (involving the diffusion of *H. heidelbergensis*), and the last starting around 200 KYA (with the origin of *H. sapiens*). The dashed lines indicate incompleteness (or lack) of the pertinent fossil record; the dotted lines suggest evolutionary relationships and/or geographic diffusions. Redrawn from Manzi (2012).

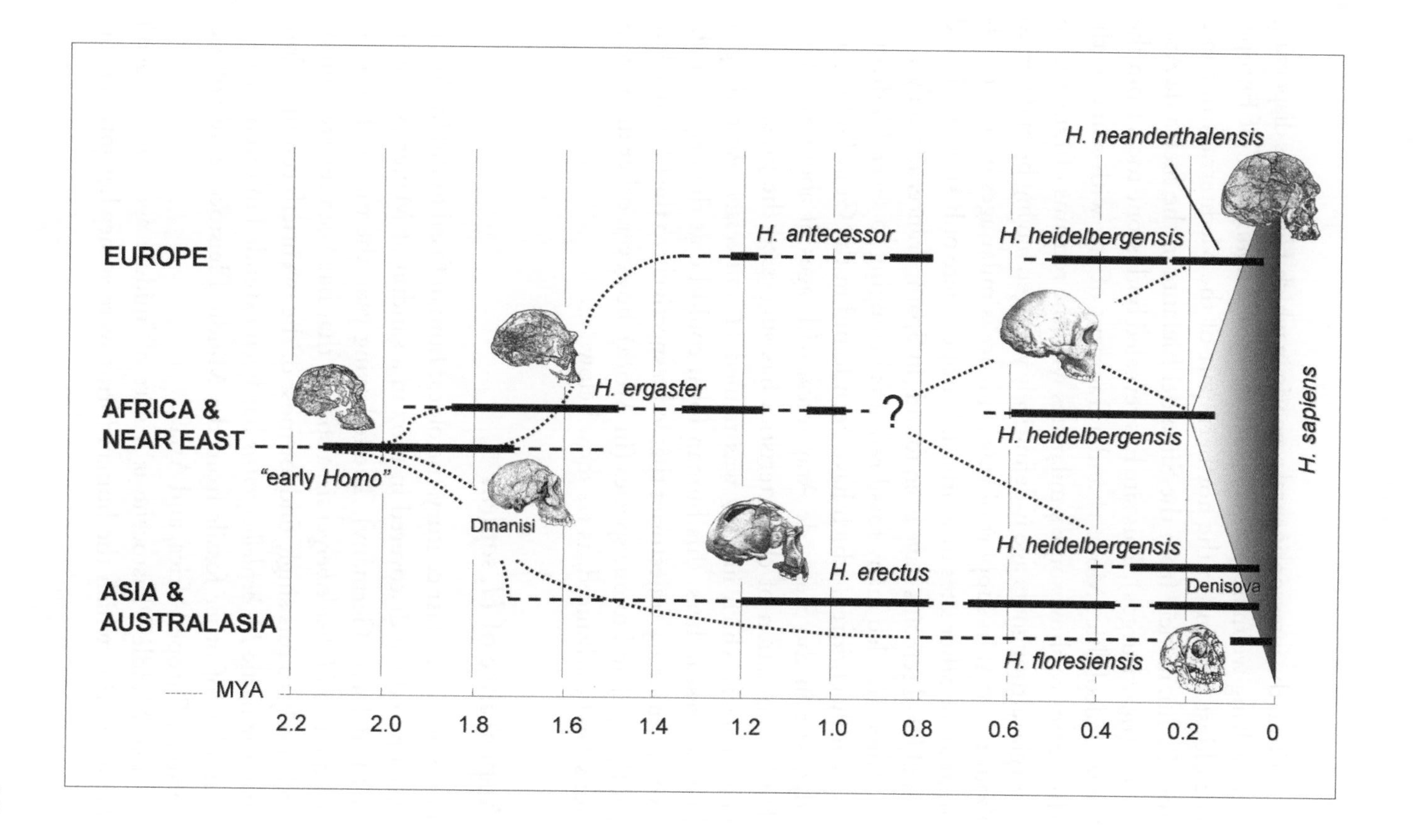

H. neanderthalensis
H. antecessor
H. heidelbergensis
EUROPE
H. ergaster
AFRICA &
NEAR EAST
?
H. heidelbergensis
H. sapiens
"early Homo"
Dmanisi
H. heidelbergensis
H. erectus
Denisova
ASIA &
AUSTRALASIA
H. floresiensis
MYA
2.2
2.0
1.8
1.6
1.4
1.2
1.0
0.8
0.6
0.4
0.2
0

Within the intercontinental scenario in which *Homo* was dispersing and evolving we must now consider the first "colonization" of Europe. The oldest remains in the northern part of the Mediterranean come from Spain, namely from the Sima del Elefante in the Sierra de Atapuerca (near Burgos). This site has revealed both stone tools from the Lower Paleolithic (Mode 1) and a fragment of jaw with some teeth. This seems to show some similarities with the remains of Dmanisi in Georgia, although its age is definitively younger, having been dated to about 1.2 MYA. In support of this evidence, similar ages have been estimated for other sites in southern Europe where only stone tools and faunal fossil remains have been found (in Spain, France, and Italy).

There are also many fossil remains belonging to several individuals, mainly children, which have been found in the Gran Dolina, another site in the Sierra de Atapuerca, with ages of about 800 KYA. This combination of characteristics has suggested the presence of a new species, which in 1997 was named *H. antecessor*. According to Spanish researchers, this human form could be at the origin of the two evolutionary trajectories that led respectively to the Neanderthals (in Europe) and to our species (in Africa); however, other interpretations can be advanced, as we show below.

Appearance of *H. sapiens*

For most of the last century, the oldest human fossil found in Europe was a mandible discovered in 1907 in a sandpit at Mauer (southeast of Heidelberg, Germany). The following year, the mandible was attributed to *H. heidelbergensis*, a species that has been referred almost exclusively to this single find for most of the twentieth century. However, the name *H. heidelbergensis* has been extended in recent years to a number of other fossils from the Middle Pleistocene, distributed between Europe, Africa, and Asia.

The Middle Pleistocene is a sort of "middle age" for the genus *Homo*, during which the human forms were somewhat intermediate between the varieties of the genus *Homo* derived from the first out-of-Africa dispersal and more recent branches of our phylogeny. As such, many of the fossil finds of this phase neither resemble any of the previous *Homo* species nor can be attributed to subsequent forms,[11] such as the Neanderthals or our own species.

The morphology of these fossil remains is still archaic, but it displays evident signs of the "encephalization" process. These humans share in fact a further endocranial expansion, accompanied by peculiar characteristics that vary from one geographical area to the other. For example, the European populations of *H. heidelbergensis* are also called "ante-Neanderthal," not only because they preceded the Neanderthals that would evolve in the same region but also because they anticipated several of the Neanderthals' morphological characteristics. This process seems to be the result of a combination of many factors: a long geographical isolation north of the Mediterranean, an adaptation to colder climates, and recurring demographic crises—with related "genetic drifts"—connected, in turn, to the glacial phases of the last half a million years.

On the other hand, the African populations of *H. heidelbergensis*, although similar in many respects to those in Europe, have no Neanderthal characteristics and, if anything, they show signs that seem to herald the emergence of our own species. At the same time, there are fossils in continental Asia that suggest the presence of human forms other than *H. erectus*, showing affinities in terms of both encephalization and other morphological details with populations of *H. heidelbergensis* found in Europe and in Africa. All this builds up the picture of a large polytypic species—*H. heidelbergensis*, precisely—that spread across several continents and existed for much of the Middle Pleistocene. The populations of *H. heidelbergensis* show differences that gave rise, over time and in different geographic and adaptive scenarios, to different species. One of these is represented by the Neanderthals (*H. neanderthalensis*).

The species *H. neanderthalensis* was distributed mainly throughout Europe, with expansions in Siberia (up to the Altai Mountains, at the border with Mongolia) and in the Middle East (including Palestine and the Indus Valley). The species appears in the fossil record around 200 KYA and disappears around 30 KYA or earlier (as we discuss in the next chapter), during the harshest period of the last Quaternary glaciation. The Neanderthals are associated with the Middle Paleolithic (called also "Mode 3" or Mousterian). This archaeological typology is characterized by new techniques of stone knapping, which, together with other lines of evidence, indicate that we are in the presence of a human form similar to ours, also in terms of intellectual capabilities and behavior. This hypothesis is also supported by the extreme levels of encephalization the Neanderthals share with modern humans, even if the morphology of their skull retained archaic features associated with other peculiar characteristics, such as the so-called chignon (of the occipital bone) or the typical mid-facial prognathism (with protruding faces around the nasal aperture).

While Neanderthals were evolving and getting established in Eurasia, *H. sapiens* appeared in Africa. The relatively recent African origin of *H. sapiens* is widely recognized, although until a few years ago it was the object of a heated debate between the proponents of this hypothesis and those who supported the idea of an ancient, multiregional origin of modern humans (a variant of the single-species paradigm). According to this model, the evolution from an archaic single species of *H. erectus* to more modern forms occurred in different regions of the world, while the gene flow among the different archaic forms and the more recent ones maintained the continuum of a single human species.

On the contrary, according to the out-of-Africa model, modern humans evolved only in Africa and replaced globally the more archaic species of the genus *Homo*. For a long time, the fossil evidence available to us—found between South Africa, Kenya, and Ethiopia—suggested that anatomically modern humans were present in Africa before anywhere else. This is confirmed by skeletons with a modern morphology found in sites in the present state of Israel, with ages close to 100 thousand years. More recently, the discovery of new fossil remains in the Middle Awash (Ethiopia) with ages of about 160 thousand years, together with the new chronology (190 KYA) attributed to a modern cranium found in the Omo Valley, have shown that in eastern Africa there were humans with fully modern anatomy at the same time as—in other parts of Africa, in Asia, and in Europe—there were humans with archaic morphologies, such as the still evolving Neanderthals and probably three other distinct *Homo* species.

A full understanding of the evolution of *H. sapiens* is a matter that will require more research, since gaps remain in the fossil record during crucial periods such as the beginning of the Middle Pleistocene, when *H. heidelbergensis* probably emerged (but we still do not know exactly when and where).

Moreover, the appearance of the modern human species is an event so important, and at the same time so "abrupt" in the fossil record, to require a very careful assessment of the genetic and evolutionary mechanisms that led to it. It is likely, however, that the crucial change intervened in a relatively small population in which the high degree of encephalization (common to different forms of the genus *Homo* in this late phase of the Pleistocene) was accompanied by a new morphological arrangement that was necessarily originated by a new regulation of the processes of growth and development. This, in turn, would have resulted in a new and more efficient balance between the cranial bones (which assume a new architecture referred to as "globular") and their contents (the brain), generating an increase in adaptive potential, especially in terms of behavioral flexibility and mental processes. This new "cultural potential" probably made a difference when the new species embarked on a dispersal that within a few tens of thousands of years led it to compete for survival with archaic human forms that were already present throughout Africa and Eurasia.

The best-known case of this coexistence is the overlap between the last Neanderthals and the first *H. sapiens,* who arrived in Europe around 40–50 KYA. These were the so-called Cro-Magnon, associated with the Upper Paleolithic (or "Mode 4") and representing a geographic expansion of *H. sapiens* from the east. The archaeological findings of material culture, including amazing artistic expressions such as rock art and music, bear witness to the appearance of a symbolic and conscious inner world, meaning that modern humans were no longer tied only to their subsistence.

These were the humans who, through trajectories sometimes discontinuous and almost never linear, would become the protagonists of a development that was not only biological, but also cultural. During their migrations, *H. sapiens* populations reached Australia (around 50–60 KYA) and later dispersed across the Bering Straits into the Americas (probably starting about 20 KYA). The timing of these dispersals and other chronological arguments will be discussed in some detail in the next chapter.

So far we have examined fossils, prehistoric sites, and artifacts. But what about genetics, or, if preferred, paleogenetics? We will deal with this topic in more detail in the final two chapters, but we can already anticipate that the current human genetic variability is limited and (on the basis of the aforementioned "molecular clock") suggests an estimated date of about 200 KYA for the origin of our species, which is definitively consistent with the fossil record. A much greater genetic variability characterizes the present African populations, and this supports the hypothesis of an African origin of *H. sapiens,* which developed from a small group of the original population of that continent.

We may add that the similarities between genetic, morphological, and archaeological data do not end here. When, in 1997, it became possible to extract "fossil" DNA from some Neanderthal remains, hypotheses emerged on both the antiquity and nature of some traits of *H. neanderthalensis* and on the estimated time of divergence of this evolutionary line from that of *H. sapiens*. In addition, there is the astonishing case of a finger bone and other apparently insignificant remains found in the Denisova cave in the Altai Mountains (Siberia), whose DNA (successfully extracted) tells the same story displayed by the fossil evidence. This is the story of the last common ancestor (*H. heidelbergensis*, who appeared around 1 MYA), of the divergence between the evolutionary trajectories of the Neanderthals and our own species (around 500 KYA), and of their speciations, which took place around 250–200 KYA, respectively.

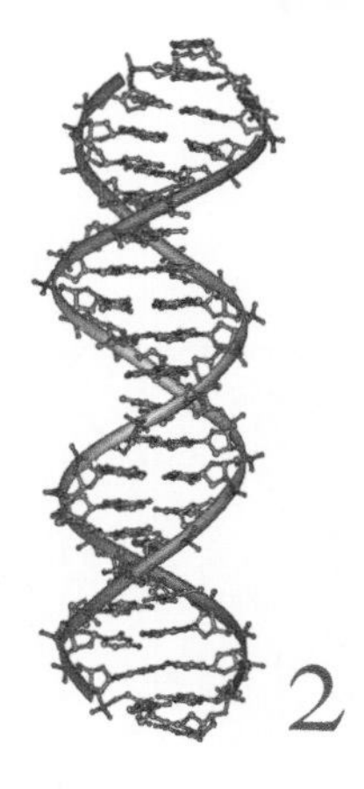

2

How (Many Millennia) Old Are You?

Measuring Time

The reconstruction of our natural history requires, as we have just seen, a precise response to a number of questions related to chronology. When did the first bipedal apes, as well as the different prehuman and human species, appear and disappear in the geological record? When did they disperse out of Africa and populate different regions on other continents? Did the extinction of the large Pleistocene animals in America and Australia or that of Neanderthals in Eurasia coincide with the arrival of *H. sapiens*? Is it possible to synchronize evolutionary events and human dispersals with geological and climatic changes? Is the human history deduced from fossil and archaeological remains chronologically compatible with that revealed by the DNA of extant humans and of some of the extinct forms? To answer these questions we need a reliable calendar.

The most accurate geologic clocks are based on radioactivity and natural radiation. In some cases we measure directly the decay of certain radioactive nuclides (i.e., radionuclides) that either were present

---

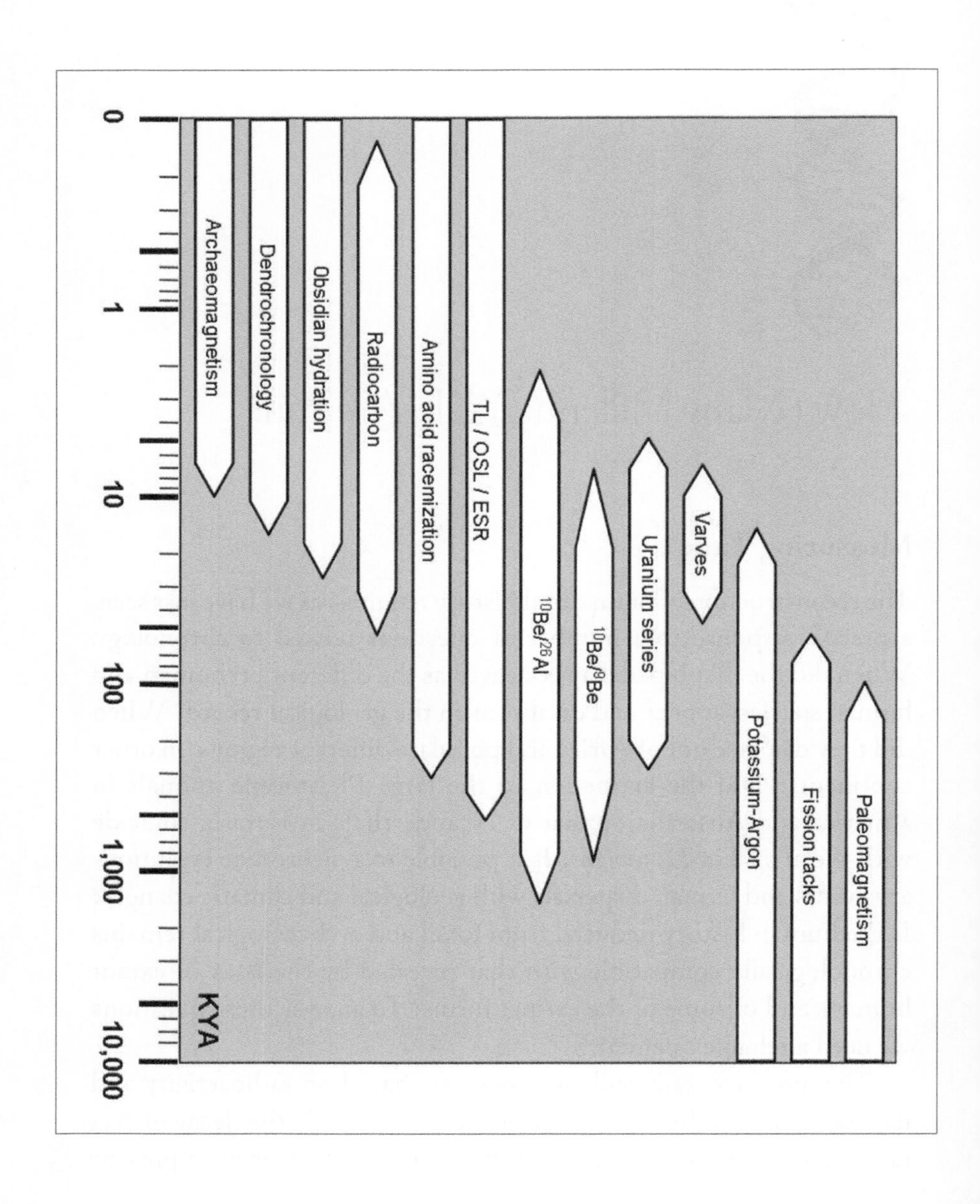

FIGURE 2.1
Time intervals covered by dating methods used in paleoanthropology.

in the Earth's crust since primordial times or were created via the bombardment of cosmic rays. In other cases ages are deduced indirectly, for example by measuring the accumulation of microscopic traces or other measurable atomic effects of natural radiation in fossilized tooth enamel or in certain crystals. In addition, we can use "relative" chronologies based on a variety of natural phenomena: sedimentation, erosion, hydration, molecular transformations, changes in the Earth's magnetic field, or the timing of plant and animal evolution.

With this in mind, we can thus begin our journey into deep time. We will review some of the most reliable techniques that are available to date archaeological and paleoanthropological findings, running our natural clocks backward. We will begin with recent case studies that are closer to our perception of time (and where our clocks can be more precise) before we move deeper into the past.

Ötzi, the Iceman

On September 19, 1991, a pair of hikers found a frozen body trapped in the Similaun glacier in the Ötztal Alps, at the border between the province of Bolzano, Italy, and the Ötztal Valley, Austria. The body was well preserved, and at first people thought it was that of a mountain climber who died recently.

Professor Konrad Spindler, director of the Innsbruck's Prehistory Institute, was the first archaeologist to see the corpse in the morgue of the city. He noted that the body was mummified, probably because of dehydration due both to the chilly winds from the north and to the föhn, a hot and dry wind that sometimes blows through the Alps. Since the shape of the axe found next to Ötzi (one of the nicknames of the mummy, after the name of the Austrian valley in which it was found) was typical of the early Bronze Age, Spindler hypothesized that these human remains should be at least 4 thousand years old. The radiocarbon method (based on the radioactive decay of carbon-14 radioisotope into nitrogen-14) was then used to date directly organic specimens taken from the body and associated objects, including clothes, weapons, and other implements.

The existence of radiocarbon in nature was discovered in 1946 by the American physicist Willard Libby, a veteran of the Manhattan Project, who used high-tech nuclear methods to measure its minute concentration in methane extracted from the sewage of Chicago. Carbon-14 should not be present in the terrestrial environment because its mean life is much shorter than the age of the Earth, which is 4.5 billion years. It is, however, continuously produced by the nuclear reactions of cosmic rays that bombard the atmosphere. Cosmogenic production balances radioactive decay, so radiocarbon concentration in the atmosphere and terrestrial environment remains fairly constant over time (we will see that this is true only as a first approximation). Carbon-14 becomes part of the food chain through respiration and photosynthesis, and its concentration in living organisms becomes approximately equal to that of the atmosphere. After the death of the organism, the concentration of carbon-14 begins to decrease (because of its radioactive decay), and its residual concentration provides the time elapsed since the formation of the living tissue.

Libby carried out his radiocarbon analyses by counting the beta particles emitted from the sample of organic material using gas detectors called "ionization chambers."[1] He was able to deduce the right age for historical materials of known antiquity, including samples from the early Egyptian Kingdoms, receiving the Nobel Prize for Chemistry in 1960 for his invention of carbon-14 dating.

Methods based on the detection of carbon-14 beta particles were improved over the years by increasing their sensitivity and precision.[2] Since the late 1970s, the accelerator mass spectrometry (AMS) method, also used to date Ötzi, has revolutionized radiocarbon dating, allowing researchers to directly count the atoms of carbon-14 instead of waiting for their slow radioactive decay. The amount of organic sample required for this type of analysis is therefore thousands of times smaller than the amount used in conventional systems, since it uses less than one milligram of carbon instead of the several grams needed with traditional methods, which must be extracted from hectograms or even kilograms of biological materials.

Small fragments of bone and tissue fibers were taken from the parts of Ötzi that were already damaged and sent to different radiocarbon laboratories. The analysis gave a very accurate estimate of its radiocarbon age—i.e., 4,546 ± 17 years BP. This age does not take into account the change over time of the original concentration of carbon-14 in the atmosphere (corresponding to zero time for the atomic clock) caused by the variability of the Earth's and Sun's magnetic fields that shield the planet from cosmic rays. In addition, the relationships between the different isotopes of carbon are affected by photosynthesis and other physical and chemical processes that determine the carbon cycle, including atmosphere and ocean circulation. To correct for all these effects, it is therefore necessary to introduce an appropriate age calibration. Dendrochronology, in particular, provides very accurate corrections for the temporal variations of radiocarbon concentration in the atmosphere.[3] The calibration is based on the measurement of carbon-14 concentration in the growth rings (of known age) of trees that are many thousands of years old. There is a problem, however: the concentration of carbon-14 in organic materials can reveal ages up to more than 50 thousand years, whereas the scale based on dendrochronology covers only the last 12 thousand years. The calibration is hence extended using sediments, corals, foraminifera, and stalactites,[4] which allow us to build a calibration curve that presently reaches to about 50 thousand years.

After the calibration, it turned out that Ötzi lived between 3,110 and 3,370 years cal BP—a measure that presents a much larger error (due to the variability in the concentration of atmospheric radiocarbon during the reference period) compared with the conventional radiocarbon age.

Much information on the mummy came from a simple visual inspection (macroscopic and microscopic). For example, the season of its death was inferred from the presence of blackberries, which ripen in autumn in the mountains, in its rudimentary backpack made of dried animal skin. In addition, pollen and wheat germ found on its clothes suggested that it came from a southern agricultural area. Its teeth were rather worn, indicating a diet of ground cereals. More information about Ötzi and the environment in which it lived came from the analysis of isotopic ratios and trace elements, a topic that will be discussed below.[5]

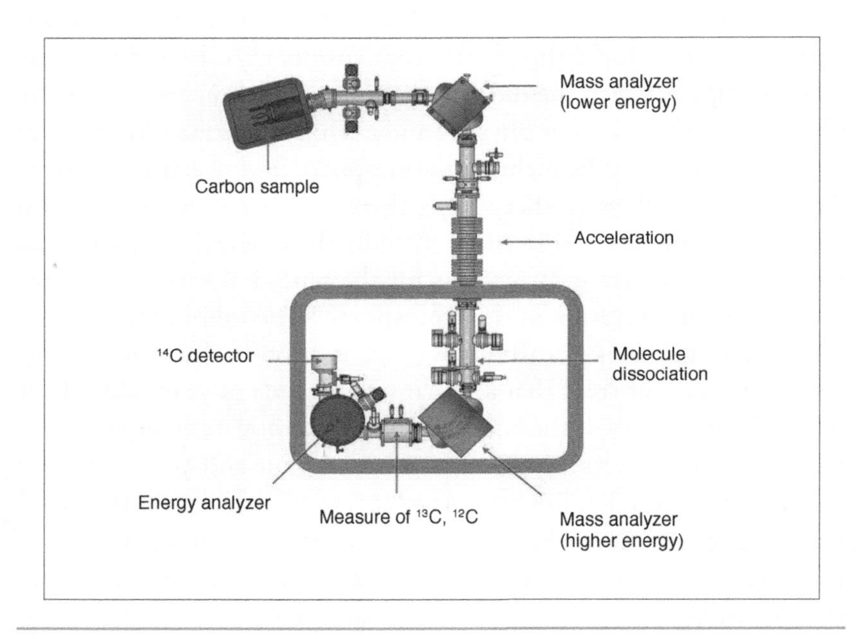

FIGURE 2.2

Miniaturized AMS system (250 kV) for radiocarbon analysis. Ionized atoms are extracted from the carbon sample, accelerated, and, after a sequence of magnetic/electrostatic analysis and molecular dissociation, identified and counted one by one in a particle detector. Isotopic sensitivities below one part per million trillion can be obtained. AMS can be also used to detect other cosmogenic isotopes, such as berillium-10. Reproduced with kind permission of National Electrostatics Corporation, USA.

In conclusion the discovery of Ötzi, a man who lived at the dawn of the Copper Age, that is relatively recently in terms of human evolution, has opened a window on the lifestyle of our ancestors. It will be increasingly difficult to access this kind of direct information when looking deeper and deeper into our past.

Master of the World

Let's turn back our clock to the previous period, classified by the archaeologists as the Neolithic, which started around 10 KYA. The entire population of *H. sapiens* at that time probably did not exceed one million individuals, but in some areas (in the Middle East, but possibly in other regions of the world, too) there were enough individuals to justify the development of agricultural practices and the domestication of animals. Recent analysis of DNA extracted from skeletons of the Neolithic found in Germany, with a radiocarbon age of more than 7 thousand years cal BP, shows that the development of agriculture in Europe coincided with the immigration of populations from the Middle East. These results trace the ancestry of the first European farmers to populations in Iraq, Syria, and neighboring countries. The spread from these regions was probably also the effect of population growth induced by the increased availability of food. From that moment on, thanks to new forms of agricultural production and animal breeding, the population of *H. sapiens* grew exponentially.

We may now have a closer look at the history of the main human dispersals and their impact on the environment. A first case in point is the relatively recent dispersal of human populations in a continent that had never been colonized by hominins for the greatest part of the human evolution history: the Americas.

Indeed, during Ötzi's time, but also during the "Neolithic Revolution," *H. sapiens* had already been in the Americas for thousands of years. There are many theories, with varying degrees of credibility, on the first human dispersal into the New World. One of the most supported theories in the past was called "Clovis First," and it assumed that North America was reached through Beringia at the end of the last glaciation by people bearing the Clovis cultural tradition. The radiocarbon dates for Clovis sites suggest that the Clovis culture lasted only 450 years, from 13,250 to 12,800 years cal BP. The Clovis First model postulates the rapid dispersal of these populations in a land without people. Their presence is evidenced by blades and bifacial points characteristic of the so-called Nenana culture discovered in Alaska, which date back to 300 years earlier than the oldest Clovis site.

In reality, the enigma surrounding the earliest Americans is complicated and can only be solved by combining data from many disciplines, such as archaeology, paleoecology, genetics, and paleoanthropology, on the basis of more precise direct dates. Thus, in the last decades, many theories on the earliest human dispersal/s in the Americas have been put forward on the basis of a growing variety of evidence, and the Clovis First hypothesis has now been abandoned. Modern humans in fact were already in Beringia 32 thousand years cal BP, as demonstrated by the presence of lithic industry and human remains in northern Siberia. Other more recent archaeological sites have revealed the presence of microblades and burins in Alaska and eastern Siberia, at least 14 thousand years cal BP.

The geological record indicates that at the time of the Last Glacial Maximum, about 21 thousand years cal BP, when the sea level was 120 meters lower than it is today, Beringia—which now consists of a chain of islands across the Bering Straits—was a vast ice-free plain with a rich variety of plant species, herds of mammals (mammoth, bison, etc.), and corresponding predators (lions, saber-toothed tigers, etc.). The populations from Beringia could not cross Alaska, however, because it was isolated from the rest of the American continent by an ice wall that stretched from the Pacific to the Atlantic coast (the Laurentide and Cordillera glaciers). Paleoclimatic studies show that the first ice-free corridor became accessible along the Pacific coast about 15 KYA, whereas a second inland corridor did not open before 13–14 KYA.

At the same time, dates of archaeological sites supporting the hypothesis of an earliest spread in America in very ancient times, before (say) 50 KYA, have been discarded by the scientific community, and several more convincing dates around 15 KYA have been obtained in the last ten years.

In 2012, a comprehensive study of the genetic diversity of Native Americans compared them with several Eurasian populations. It turned out that the group of modern humans from whom most of today's Native Americans have descended crossed the Bering Strait more than 15 KYA. It seems that two further human waves followed, from which the first peoples of the Arctic and some populations of native Canadians have descended.

Genetic data—combined with anthropological, archaeological, and paleoenvironmental data—also reveal details of other migrations of early modern humans from Central Asia and, originally, from Africa. For instance, evidence suggests that humanity faced a bottleneck around 60–70 KYA, with the effective population size of *H. sapiens* declining to a few thousand individuals. This event has been linked to the environmental disaster caused by the Toba volcano that occurred in Sumatra 74 KYA. The eruption of the Toba volcano was the most violent of the last 2 million years, pouring 2,700 cubic kilometers of ash into the atmosphere and causing long-term climatic changes. The disaster dramatically increased the dryness of the African regions, probably forcing modern humans to move to refugia on the shores of eastern and southern Africa and to develop new skills in coastal navigation. Some human groups left the continent, probably following the movement of other animals.

Saudi Arabia is a key area for the reconstruction of the dispersal of modern humans from Africa (there is no agreement whether they were only "anatomically" modern or also "behaviorally" modern, as we shall discuss later). Given the difficult environment that now characterizes Saudi Arabia, the question of the chronology of human dispersal is also linked to paleoclimate, since the migration routes were influenced by the availability of food and drinking water. The deposits left by ancient lakes in the geological stratigraphy of some Arabian regions were dated with optically stimulated luminescence (OSL), a dating method based on the accumulation of energy in certain sediment crystals as a result of natural ionizing radiation.

The relict lakes of Saudi Arabia have shown OSL ages of about 80, 100, and 125 thousand years. During these periods, therefore, the environmental conditions in this area were favorable to human migrations. Starting 75 KYA, the Toba-induced arid climate transformed Saudi Arabia into an insurmountable barrier. Thus, the dispersal of modern humans along the southern corridor that led from the Horn of Africa to Asia must have occurred before 75 KYA, probably in successive waves. This hypothesis is supported by archaeological studies. Stone tools attributed to the Middle Paleolithic have been found in the Nefud desert in northern Arabia, associated with geologic deposits located near a relict lake that have been dated with OSL to 75 KYA. Stone tools characteristic of the Middle Paleolithic in Africa have also been found at Jebel Faya, United Arab Emirates, demonstrating the presence of *H. sapiens* in eastern Arabia during the last interglacial period.

Until recently most scientists believed that modern humans, after leaving Africa about 60 KYA, dispersed along the coasts of the Indian Ocean following a single rapid event. The coastal route would then take them all the way to Australasia. According to this hypothesis, the new "migrants" carried microlithic technologies and around 35–45 KYA replaced the existing populations of southern Asia, associated with Middle Paleolithic industry.

However, on the basis of new archaeological, genetic, and paleoenvironmental data, including more precise direct dates, a growing number of scientists are supporting the hypothesis that modern humans were in Arabia and southern Asia much earlier, possibly between 130 and 70 KYA, and probably at the same time as the *H. sapiens* was in the Levant. According to this model, modern humans left Africa carrying technologies associated with the Middle Paleolithic and Mesolithic. About 35–40 KYA these industries would then evolve in situ in southern Asia, giving rise to microlithic industries. Long archaeological sequences of human occupation have been found at Jwalpuram, in Andhra Pradesh, India, supporting this transition. Other evidence comes from the archaeological remains discovered in the valley of Jurreru (in Kornool, India), which shows continuity of the local Middle Paleolithic industry up to 38 KYA. These technological innovations, according to paleoanthropologists, would have originated in response to either population growth or environmental deterioration (increased aridity approaching the Last Glacial Maximum, 21 KYA) in South Asia.

Thus, human dispersal in Arabia and South Asia and its demographic fluctuations seem to be much more complex than has been assumed so far. Sites with lithic industries characteristic of the Middle Paleolithic were found in southern India before and after the eruption of Toba in strata dated with high precision using the OSL method. In this area, although the environmental situation had already deteriorated before the volcanic event, the changing conditions did not interrupt the continuity of human presence.

India's role in the dispersal of modern humans can be inferred from the genetic study of its indigenous peoples (see chapter 5). It has recently been suggested that the Solinga, who live in the Bilingiri Rangana hills in southern India, could have descended from the early modern humans who inhabited the subcontinent. They are different from all other Indian populations, but they show an incredible genetic affinity with two Australian Aboriginal populations. One could therefore believe that the Solinga are connected to the first migration wave along the southern Indian coast that eventually reached Sahul; but such a genetic affinity could also be explained by a more recent dispersal from the Indian subcontinent to Australia.

Remaining approximately in the same chronological horizon of the final stage of the Pleistocene, we now move on to Sahul.

Landfall in Sahul

The oldest human bones found in Australia date back to at least 40 KYA, as evidenced by the ancient human bones found in the Willandra Lakes (presently a UNESCO World Heritage Area), 800 kilometers west of Sydney. It was here that in 1969 Australian archaeologists discovered the remains of the Mungo Lady, a Pleistocenic woman who died at about twenty years of age. The way in which the ancient inhabitants of Mungo prepared the woman's body for her funeral ceremony is extraordinary. After cremating her, they removed the skeleton from the ashes and crushed it, paying particular attention to the skull, before sprinkling the fragments with powdered ochre and burying them. The first radiocarbon dating of the Mungo Lady gave an age of 29 thousand years. Further radiocarbon analyses, however, indicated an older age of about 40 thousand years.

In the following years the bones of other individuals, including the skeleton of an adult male and a child, were found in the same area. The age of the Mungo Man has been estimated between 60 and 74 thousand years using uranium series dating through the nondestructive analysis of the sample's gamma radioactivity. To obtain these estimates the skull fragment, weighing 305 grams, was placed in a lead-shielded chamber equipped with detectors to count the high-energy gamma radiation emitted from the minute amounts of natural uranium contained in the fossilized material.

Further analyses were carried out in 1999 using the electron spin resonance (ESR) method. ESR is based on the effects of naturally radioactive elements, especially uranium, on the electronic structure of crystals (e.g., on tooth enamel). The ESR analysis gave an age of more than 60 thousand years for the Mungo skeletons. It is apparent, however, that the dates obtained directly from the Mungo bones are not entirely congruent with the OSL dates of the sediments in which they have been found. Using the OSL technique, the sediments of the Mungo burials were dated to 38–42 KYA, which is compatible only with the latest age attributed to the remains of the Mungo Lady and does not match the date of more than 60 KYA attri-buted to the bones of the Mungo Man. However, other data support the hypothesis that modern humans were already in Australia 60 KYA, also based on OSL dating of archaeological sites (containing only stone artifacts) found in Malakunanja and Nauwalabila, in the north of the Australian continent. Because of these inconsistencies, the question of when humans first reached Australia remains open and controversial.

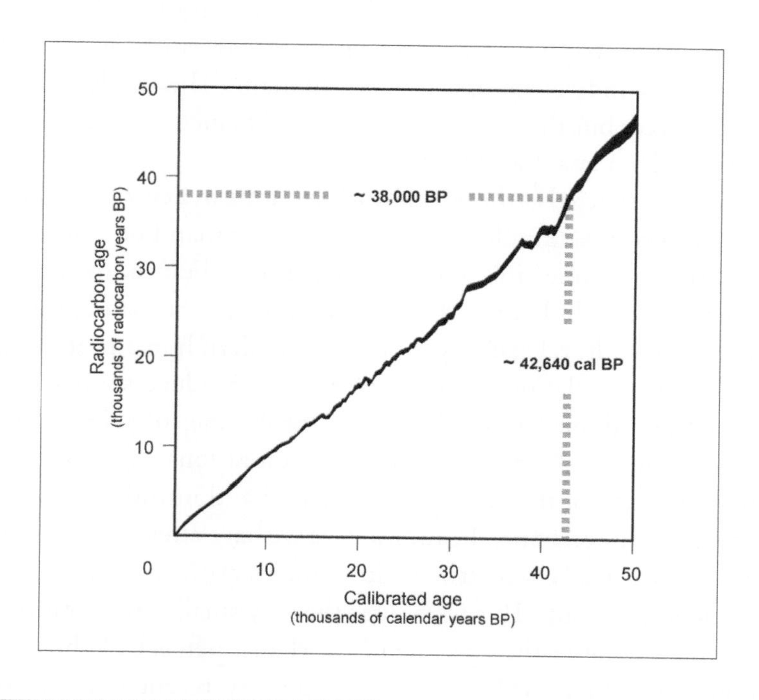

FIGURE 2.3
Calibration curve to convert radiocarbon ages into calendar ages. From Reimer et al. (2009).

H. Sapiens Meets Distant Relatives

Radiocarbon dating is the most precise method that can be applied to evaluate the period in which early populations of *H. sapiens* overlapped with the Neanderthals in Europe. Some years ago, Paul Mellars from Cambridge University analyzed all the AMS radiocarbon dates to reconstruct the chronology of the presence of Neanderthals and modern humans in Eurasia. Using appropriate calibration corrections, he showed that the overlap was close to 5 thousand years and not 10 thousand years, as previously assumed. Other analyses suggest an even shorter temporal overlap between the first modern Europeans and the Neanderthals. The Oxford Radiocarbon group has recently dated more than 500 organic samples from seventy Paleolithic sites throughout Europe characterized by lithic industries attributed to the Neanderthals and to the first modern humans. The results are still being analyzed, but they hint to a fast disappearance of the Neanderthals, possibly in less than 2 thousand years.

Studies conducted at the Max Planck Institute in Leipzig, Germany, have used direct AMS radiocarbon dating on human bones found in the Grotte du Renne sites and Saint Césaire in France. As a result, the scientists concluded that between 40 and 50 thousand years cal BP there were cultural exchanges between modern humans and some Neanderthal populations, who were able to produce sophisticated bone tools, small blades, and body ornaments: the so-called Châtelperronian industry. Unfortunately, the calibrations of these dates stretch the limits of the radiocarbon method and introduce a significant degree of uncertainty. In addition, as we have already mentioned, particular care needs to be taken with samples exceeding an age of 40 thousand years cal BP. This is because the very small concentration of carbon-14 residue in the specimen (less than 1 percent of the initial concentration) may easily be contaminated by recent radiocarbon-rich organic materials both in the field and during laboratory preparation (e.g., during the various processes of combustion, hydrolysis, and graphitization). Recent discoveries in France, dated with both radiocarbon and OSL to 51 KYA, suggest that Neanderthals created the first specialized bone tools in Europe. The ribs of reindeer or similar animals found in the Mousterian site of Pech-de-l'Azé were shaped to the desired form by grinding them against a coarse material, a technique that perhaps influenced modern humans of the later Upper Paleolithic at the time of their arrival in Europe.

In any case, around 30 KYA (but this date is still being debated) the Neanderthals disappeared throughout Eurasia along a route that goes from east to west. This dispersal was apparently correlated to the dispersal of *H. sapiens* populations in Europe, as evidenced by cultural manifestations—stone and bone artifacts, rock art, musical instruments, etc.—that in fact moved from east to west approximately during the same period. The products of modern humans such as bone and ivory flutes and female figurines, dated by radiocarbon to 40–45 thousand years cal BP, were recently discovered in the Hohe Fels Cave in southern Germany and were assigned by archaeologists to the Aurignacian culture. The remains of the last Neanderthals, not surprisingly, were found in Gibraltar caves, with a date of 28 thousand years cal BP. This date is now considered to be an underestimate after more advanced radiocarbon techniques (based on better pretreatment procedures to avoid contamination of the organic samples) carried out in 2013 provided much older dates, around 50 thousand years cal BP, for sites in southern Spain that had previously been dated to around 35 thousand years cal BP.

During their dispersal in the opposite direction, toward Asia, modern humans could have met the last representatives of *H. erectus* as well as the recently discovered humans, presently known only from the scanty (but very informative) remains of the Denisova cave, or even our weird cousin, *H. floresiensis* (already introduced in the previous chapter).

The grains of quartz and feldspar collected in the sediments associated with the *H. floresiensis* remains have been OSL-dated to less than 30 KYA, whereas radiocarbon dating returned an age of about 18 thousand years cal BP for small pieces of charcoal found in the same material. In subsequent years, a research team has collected the remains of at least six individuals of the same species that have been dated to between 20 and 12 KYA; associated stone tools cover a period that ranges from 95 to 12 KYA. The most recent dates are presently being revised. The Liang Bua lithic industry resembled the artifacts from Mata Menge, an archaeological site in the center of the island, where in 1998 Morwood and his colleagues had obtained an age of over 840 thousand years using fission track dating. Another site, dated using argon-40/argon-39, revealed the presence on Flores of stone artifacts one million years old.

The evolutionary picture emerging from Flores looks complex and its interpretation is not shared by all researchers, though the details of this debate will not be discussed here. What we would like to emphasize is that in Liang Bua almost all the available methods of geochronology were applied, from radiocarbon to OSL, uranium series, and ESR, using a range of materials, including coal, feldspar, calcium carbonate, and tooth enamel.

"Modern" Behaviors

Until recently, conventional wisdom held that modern human behavior emerged in Europe during the Upper Paleolithic, when impressive cultural manifestations appeared in the archaeological record. Radiocarbon-dated archaeological sites show that "suddenly," between 45 and 35 thousand years cal BP, *H. sapiens* felt the need to invest a considerable effort in activities that went beyond the basic needs of survival and began the extensive use of ornaments, pictorial representations, symbols, and musical instruments, as we mentioned above. Of course, these culturally modern humans also perfected the stone tools they needed for their survival (for hunting, butchering, processing wood and other materials, etc.). They also made tools of growing complexity using bone, ivory, wood, and other materials.

However, a growing body of evidence shows that such modern behaviors originally developed in Africa, anticipating by tens of thousands of years the Upper Paleolithic "revolution" in Europe. Here we are really in the deep past, beyond the reach of radiocarbon techniques. To settle the question, other methods, including OSL dating, can be used.

In the Blombos Cave in South Africa, in sediments dated by OSL to 70–80 KYA, archaeologists found seashells of the species *Nassarius kraussianus* with artificial perforations and pieces of ochre with graphic symbols. They also found in the same region advanced stone tools, the so-called Still Bay points, which suggest innovative techniques unknown in previous periods. Another archaeological lithic industry, called Howieson Port, provides further evidence of human innovation in this area. The OSL dates show that the Still Bay technology appeared 71 KYA and lasted only a thousand years, whereas the Howieson Port industry lasted five millennia, from 65 to 60 KYA. Archaeological studies and population genetics, which will be discussed in the final chapters, reveal that these periods of technological innovation in Africa corresponded to increases in population. In 2012, it was discovered that at Pinnacle Point, on the southern coast of South Africa, advanced lithic technology (microlithic) had not been so sporadic but had evolved continuously for a period of 11 thousand years, starting around 71 KYA. OSL provided a precise chronology for a 14-meter sequence of archaeological sediments covering the period from 90 to 50 KYA. According to many scholars, these innovations would be convergent with similar technological developments found in southern Asia 35 KYA, as we discussed earlier.

It seems that the cognitive leap of the Upper Paleolithic incubated in the minds of early *H. sapiens* in Africa for more than 100 thousand years, while the world around them was changing due to global climate disruptions and in synchrony with the appearance and extinction of various animal species. During this period, the lithic technology evolved very slowly, and symbolic manifestations were totally absent. But when and where did the anatomically modern humans appear?

The Earliest "Modern" Humans

In 1932, paleoanthropologist T. F. Dreyer discovered at Florisbad, in South Africa, the incomplete remains of a human cranium that seemed to belong to *H. sapiens,* although the specimen had some archaic characteristics, including a receding forehead and a developed supraorbital torus. Half a century later, in 1974, conventional radiocarbon dating was performed on the peat considered contemporary with the human remains and returned an age of about 40 thousand years. This estimate was confirmed by subsequent dates obtained with amino acid racemization (AAR, a dating method based on the chemical induction of slow changes in the amino acids of organic materials) using the tooth of a hippopotamus that was also considered coeval with the cranium—a sign that the area, now dry and infertile, was then rich in lakes and rivers.

The above dates might confirm the hypothesis that the Florisbad cranium belonged to humans with archaic features who survived in Africa until relatively recent times, when they were likely replaced by fully modern humans arriving from the north. But another geological clock, which could be applied directly to the only available tooth, a third molar of the Florisbad man, returned surprising results. Geocronologist Rainer Grün from the Australian National University, using the ESR technique on a small fragment of enamel, obtained a date of 259 ± 35 thousand years. Hence, the Florisbad cranium belonged to one of the last archaic humans, showing features that anticipated the emergence of modern forms: one of the last, indeed, since the first *H. sapiens* would emerge shortly afterward. Similar cases are represented by various African remains from the late Middle Pleistocene, such as the remains of Guomde in Kenya, which included a partial cranium and a femur that were dated to 272 KYA by the uranium series method, using gamma spectrometry.

The remains of the earliest anatomically modern humans have been discovered in Ethiopia: Omo Kibish 1 and Herto. The skull of Omo Kibish 1, discovered in the valley of the Omo River, immediately north of Lake Turkana, is currently the oldest known *Homo* fossil to show a fully modern morphology, characterized by a rounded braincase, a protruding chin, and the lack of a supraorbital torus (even if the supraorbital ridges are quite robust). Argon-40/argon-39 dating, a method based on the radioactive decay of potassium-40 into argon-40, has recently provided an age of 196 thousand years for the volcanic ash located immediately below the site where this skull was found.

Found farther north, in the Middle Awash region of Ethiopia, the Herto remains include the skulls of two adults (with a general morphology that seems rather modern, albeit with some robust and archaic features) and a child (in which the modern morphology is even more evident). At the time of their discovery, the skulls were sandwiched between two layers of volcanic ash that have been dated to 154 and 160 KYA using the same argon-40/argon-39 method. Acheulean and Middle Paleolithic stone tools were also found at these sites. As we said before, such stone tools are representative of the culture of the early *H. sapiens*, which will not show more complex manifestations for at least another 100 thousand years.

Around 120 KYA, the descendants of these early modern humans left Africa and dispersed into the Levant, where they probably encountered *H. neanderthalensis* for the first time. Decades of studies using all possible geological clocks, from uranium series to ESR and luminescence, have been applied to a range of materials—such as faunal teeth and burnt flint—extracted from the Skhul cave of Mount Carmel and the cave of Qafzeh near Nazareth, where archaeologists found several skeletons of modern-looking humans. It is presently suggested that the ages of these remains range from 120 to 90 thousand years. In other caves of Mount Carmel, such as Tabun and Kebara, as well as in a cave just north of Lake Tiberias (Amud), all yielding Neanderthal fossils, similar dating methods show that also this human species arrived in the region after 120 KYA, to disappear 60 KYA.

The debate is still open, but we can confidently assume that the Palestine region was alternately occupied for many tens of thousands of years by both *H. sapiens* and *H. neanderthalensis,* as a result of the dispersal of each of these human species in sync with climate change. Our ancestors were dispersing northward during the last interglacial period, 120 KYA, whereas Neanderthals abandoned the arid and cold Europe in the following Ice Age to take refuge in the environmentally more attractive eastern Mediterranean lands.

But who were the ancestors of these two species? How can we determine the timing of the complex scenario represented by the evolution and dispersal of the genus *Homo*?

Big Bang *Homo*

To have a look at *Homo* as a diverse and cosmopolitan genus, we must move further back into the past. By the end of the Lower Pleistocene, around one million years ago, archaic human groups had inhabited much of Africa, the Caucasus, and Southeast Asia for an extended period, and they had finally dispersed into Eurasia for its entire length east to west, from the island of Flores in Indonesia to Europe's Iberian Peninsula.

But which is the oldest human fossil found in Europe? And who were the first Europeans? A jawbone found in the archaeological site of Sima del Elefante, in Spain, is about 1.2 million years old, as determined through paleomagnetism and age estimates based on animal fossils of known antiquity (biochronology), especially the so-called micro-mammals (mostly rodents). This chronology was recently confirmed by cosmogenic radionuclide dating (using beryllium-10/aluminum-26 dating, a method based on the decay of the radionuclides beryllium-10 and aluminium-26, which are produced by cosmic rays on the Earth's surface), corroborating the oldest evidence of human presence in Europe. As anticipated in the previous chapter, the mandible from Sima del Elefante may belong to an intermediate form between the species *H. georgicus* (from Dmanisi, Georgia) and the species *H. antecessor*, which is based on a large sample of fossils from Gran Dolina, another site in the Sierra de Atapuerca, Spain.

Moving further back in time, we find the most ancient *Homo* species that is widely recognized by paleoanthropologists: the African *H. ergaster*, associated (it is not clear so far whether at the interspecific or intraspecific level) with its Asian variant, *H. erectus*. The most complete cranium of *H. ergaster* is KNM-ER 3733, discovered in the Koobi Fora formation, Kenya, in 1975. The cranium, which belonged to a female with a brain capacity of 850 cubic centimeters, was dated to 1.78 MYA by applying argon-40/argon-39 to the volcanic ash covering the geological layer that included the cranium. Other fossil remains of *H. ergaster* in Koobi Fora (cranial fragments and bones of the coccyx) have an age of 1.9 million years. *H. ergaster* fossils were also found at Swartkrans in South Africa with an age of 1.8 to 1.9 million years, determined by applying uranium-lead dating to the concretions of calcium carbonate that contained the fossil remains. This method is based on the analysis of the decay of uranium-235 into lead-207 ($T_{1/2}$ = 704 thousand years) and of uranium-238 into lead-206 ($T_{1/2}$ = 4,468 thousand years).

H. georgicus has been dated to 1.8 MYA with argon-40/argon-39 using samples from the underlying basalt formation. This chronology is confirmed by the presence of a geomagnetic excursion corresponding to this geological period (the Olduvai subchron, 1.85 to 1.78 MYA). The fossil evidence from Dmanisi may even belong to the first *Homo* specimen to disperse out of Africa, well before Asian *H. erectus* varieties such as Java Man and Peking Man.

While it is generally demonstrated that *H. ergaster* existed from at least 1.9 MYA, the status of *H. habilis* and *H. rudolfensis* is much more controversial. Skeletal remains assigned to *H. habilis* have ages ranging from 2.33 to 1.44 million years. A jaw, AL 666-1, found at Hadar in

Ethiopia in 1994, is the oldest known hominin fossil that is attributed to *H. habilis*. It was found in a geological horizon rich in stone tools, 80 centimeters beneath a layer of volcanic tephra, which was dated using argon-40/argon-39 to 2.33 ± 0.07 MYA. In 2000, still in the Koobi Fora formation, east of Lake Turkana, paleoanthropologists found the right part of a jaw, KNM-ER 42703, later assigned to *H. habilis*, with an age estimate of about 1.44 MYA determined with the argon-40/argon-39 method applied to the surrounding tuff. Stone tools associated with *H. habilis* were found at Hadar in Ethiopia in sediments dated to around 2.5 MYA. The age was obtained by various methods, including fission track dating, based on the natural fission of uranium-238 in volcanic mineral.

As we said, and as far as we know, *H. habilis* had long arms, and its physical dimensions were similar to those of *Australopithecus*, making it difficult to distinguish the two genera in this first phase of existence of the genus *Homo* (if we may definitively consider *H. habilis* to be *Homo*). The discovery of *A. sediba* has reignited the debate on this point, and reliable dates are needed to confirm the chronology of such a new species. The age of *A. sediba* was determined with high accuracy by combining uranium-lead dates with paleomagnetic data (based on the presence of a very brief geomagnetic excursion, the so-called Pre-Olduvai event). The age of *A. sediba* is 1,977 million years with an error of ± 2 thousand years.

Australopithecus spans an archaeological record of more than 4 million years, according to argon-40/argon-39 dating based on volcanic material found close to the fossil remains (only applicable to the East African sites). This method has been greatly improved in recent years. In particular, a laser microprobe that allows the analysis of microscopic specimens can now be used to extract argon from single crystals. For example, grains with dimensions of about 0.5 millimeters were taken from the volcanic ash lying just beneath the geological layer in which Lucy was discovered, returning an age of 3.2 million years.

As we reported in the previous chapter, *Australopithecus* was preceded by other bipedal apes, such as *Ardipithecus, Orrorin,* and *Sahelanthropus,* in order of increasing antiquity.

Here we will just discuss the chronology of *Sahelanthropus,* considered the oldest representative of the group. Unfortunately, the location where it was found, in Chad, is an area where there are no volcanic deposits, so the argon-40/argon-39 method could not be applied. An age of about 7 million years has been determined by biochronological methods using mammal fossils of known age found in the same geologic horizons as the cranium. *Sahelanthropus* has been also dated with another, less accurate, procedure: the beryllium-10/beryllium-9 method (a dating method based on the decay of beryllium-10, which is produced by high-energy cosmic rays and transferred to the Earth's surface by precipitation).

The two layers of sediments that incorporated *Sahelanthropus* have dates of 6.83 and 7.12 million years, confirming the previous estimate. However, the classification of this ancient primate of tentative age (additional cautions also exist, not discussed here) remains uncertain. We need new science and new fossils to reveal more precisely the story at the frontier(s) of the human lineage.

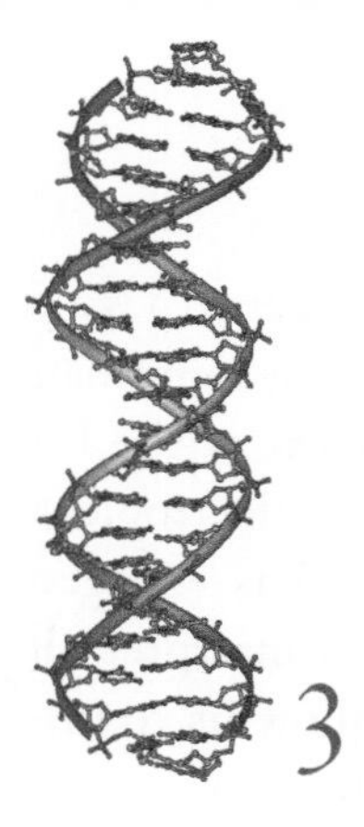

3

What Bad Weather in the Pleistocene!

Paleoenvironments

In 2012, scientists from the Soil Cryology Laboratory in Moscow were able to germinate seeds of *Silene stenophyilla* that had remained permanently frozen at −7°C for 32 thousand years. The seeds were found in the tundra of northeastern Siberia, in burrows of Ice Age squirrels near what are now the banks of the Kolyma River. The burrows were nearly 40 meters below the surface, at the same level where archaeologists had found remains of various (now extinct) Ice Age mammals. The DNA of the plants had been preserved intact, and the researchers were able to produce new plants with petals slightly different from the existing variants of the same species due to mutations that had occurred in the meantime.

During the Ice Age, *S. stenophylla* grew extensively across the Siberian tundra, which was inhabited by herds of mammoths and woolly rhinoceros, coveted prey of *H. sapiens* hunters. At the same time, the last Neanderthals were still surviving in western Europe, and there were even other species of *Homo* distributed along the edges of the

Asian continent. The discovery by the Russian scholars opened a window onto the environment of a period of extreme interest in human history, with different populations subjected to the stress of the Ice Age (freezing cold in the north accompanied by drought in the tropics).

Past climates and environments are a puzzle that has engaged generations of scientists. The goal has always been to interpret all the clues, in the biosphere and the lithosphere, that allow us to understand the landscapes frequented by thousands of human generations before us and, earlier, by the hominins from which our species evolved. Since it is not easy to find remains such as these Siberian seeds, which were still able to germinate, we need to use other indicators of past environmental conditions.

Pollen, for example, can be preserved in the sediments of ancient lakes for hundreds of thousands of years. Knowing the chronology of a sediment core, it is possible to reconstruct environmental changes in the deep past. There are other environmental indicators, too, based on isotopic fractionation: certain biogeochemical processes, such as evaporation, photosynthesis, and the precipitation of minerals, select specific isotopes of an element and change their relative abundance—for example, the abundance of oxygen-18 relative to oxygen-16, that of hydrogen-2 relative to hydrogen-1, or that of carbon-13 relative to carbon-12. The isotopes of oxygen and hydrogen allow us to evaluate past temperatures in marine sediments, Antarctic ice, stalactites, and other natural archives in which such information is stored for hundreds of thousands and even millions of years. On the other hand, the concentrations of stable isotopes of carbon can provide information on the carbon cycle and hence on atmospheric and oceanic circulation, which are critical in the dynamics that determine climate evolution.

The isotopes tell us that around 50 MYA the global temperature began to fall, slowly bringing the Earth from the tropical conditions that characterized the beginning of the Cenozoic to the ice ages of the Quaternary. The temperature decrease during the Cenozoic was accompanied by a CO_2 decrease in the atmosphere, from 2,000 to 300 parts per million. Some argue that this was the main cause of the global cooling. Others suggest that the harsher climate was due to changes in ocean circulation produced by the movement of tectonic plates. More probably, these and other factors overlapped, getting amplified by complex feedback phenomena. The environmental archives reveal that the cooling process was interrupted by periods of temperature increase, after which the cooling trend would start again. The isotope analyses of marine sediments suggest that the temperature of deep ocean waters decreased from the 12°C of 50 MYA to 6°C around 30 MYA (today this temperature is 2°C).

During the Cenozoic the supercontinent Gondwana split, Antarctica moved to its current location at the South Pole, and South America and North America joined together. Of course, these changes in the lithosphere dramatically influenced the oceanic currents and atmospheric circulation, which play a crucial role in the distribution of the heat coming from the Sun. Then, around 40 MYA, the Indian plate collided with the rest of the Eurasian continent, forming the high peaks of the Himalayas and raising the Tibetan plateau; this barrier changed the atmospheric circulation, and its newly formed rocks absorbed large amounts of CO_2. Lower CO_2 concentrations reduced the greenhouse effect. Combined with changes in oceanic currents, global temperatures plummeted, causing the environment to become more arid in many parts of the world.

As we mentioned in the previous chapter, the first apes—or, better, hominoids—appeared almost 30 MYA among the African primates, when Africa was completely severed from the other continents. At the beginning of the Miocene, after the displacement of the Arabian plate, Africa and Eurasia touched again. Hominoids were thus able to expand over an area characterized by dense forests, which stretched across Africa, and then from the European shores of the Atlantic Ocean to the Asian shores of the Indian Ocean. For several million years, a wide variety of hominoids (over 100 species) lived in the vast lands of this supercontinent, with the sea (an ocean called

Tethys) periodically separating and reuniting Africa and Eurasia. Among these hominoids were the ancestors of extant apes, including our own species.

The global climate then remained relatively constant until about 10 MYA, when the cooling trend restarted. The global temperature decrease caused an extension of the arid areas. Toward the end of the Miocene, around 5 MYA, the forests that covered vast parts of Africa and Eurasia had been broken into a mosaic landscape that included both wooded (forest) and more open grassy areas. In Africa, tectonic events contributed to the climate upheaval in the eastern part of the continent. The separation of two of the African plates created the GRV, a giant trench 6,000 kilometers long, which begins in southern Mozambique, crosses eastern Africa, and carries on northward up to the Red Sea and the present Middle East. What was a relatively flat area covered with forest became a landscape of mountains, lakes, cliffs, and ravines. The area corresponding to present-day Ethiopia rose by more than 2,000 meters. A similar uplifting occurred in Kenya. High mountains rose in the west in a direction parallel to the GRV, while along the trench vast river and lake basins opened up, some of which still exist. The new landscape blocked the wet perturbations coming from the Atlantic Ocean, favoring dry monsoons from Asia that increased dramatically the drought in this part of Africa.

This was the landscape that favored the evolution of our ancestors. Let's recall that after the appearance of bipedalism—a sort of precondition, in turn strongly influenced by environmental changes—the main milestones in human evolution have been (a) the emergence of the genus *Homo*, to which we associate the first appearance of Paleolithic stone artifacts; (b) the increase in brain size (or encephalization), together with important changes in the life history[1] of our ancestors; (c) a series of geographical dispersals, mostly from the African continent to Eurasia, and the emergence of a considerable number of species belonging to the genus *Homo*; and (d) the origin of *H. sapiens* and the development of modern human behavior. Of course, this set of events is not part of a linear process: the evolution and worldwide dispersal of *H. sapiens* were neither predetermined nor inevitable. The geological archives tell us unequivocally, however, that many of these evolutionary steps coincided with important environmental changes in Africa and Eurasia, giving increasing support to the idea that the environment influenced the biological evolution and behavior of humans in a profound and decisive way.

Once upon a Time, the Ice Ages

The Pleistocene was characterized by climatic fluctuations more intense and more frequent than those of the previous period. As we have seen, it is at the very beginning of this epoch that the genus *Homo* appeared, together with the first evidence of Paleolithic stone tools in eastern Africa. The earliest representatives of *Homo* had a slightly larger brain (on average) than *Australopithecus*; they were more accomplished bipeds and began dispersing into other parts of Africa and Eurasia. The adaptation of these archaic humans was primarily biological, even if their technology (first Oldowan, then Acheulean) contributed to their survival in environments beyond the tropical areas and under the new climatic conditions of the Pleistocene. For the first time *Homo* met with extreme global temperature fluctuations corresponding to alternating glacial and interglacial phases. We now live in the Holocene epoch, considered the latest interglacial period.

During the ice ages, large amounts of water accumulated on dry land, forming glaciers that periodically covered extended areas of the northern hemisphere and Antarctica. The periodic melting of this ice during the interglacial phases raised the sea levels. Even back in the nineteenth century, geologists had identified the glacial origin of the boulders located more than hundreds of kilometers away from the nearest areas of bedrock outcrop. They were studied in relation to the glaciers of the European Alps, refuting the idea that the huge rocks had been transported by the waters of the Biblical Great Flood. Swiss geologist Louis Agassiz was one of the first to explain scientifically these "erratic" boulders and the effects of their displacement, linking them to the ice ages.[2] Later he developed the idea that the ice ages could be related to the orbit of our planet around the Sun. Initially, it was assumed that global temperatures decreased when winters—corresponding to the maximum distance between the Earth and the Sun (aphelion)—lasted longer, providing a weaker solar irradiation.

In the 1930s, Milutin Milankovitch (a Serbian mathematician, 1879–1958) assumed instead that the ice ages were caused by a reduction of solar energy in the northern hemisphere during the summer months. This situation would favor the preservation of glaciers, causing them to accumulate into increasingly large ice caps. In Milankovitch's model, the ice ages that occurred periodically on our planet could be attributed to changes in the three parameters that describe the Earth's orbit around the Sun: obliquity (the tilt of the planet with respect to the orbital plane), eccentricity (the deviation of its orbit from a circle), and precession of the equinoxes (the oscillation of the orbit's axis). According to the Milankovitch theory, obliquity, eccentricity, and precession vary cyclically, at intervals of 41, 100, and 23 thousand years, respectively. Together, these cycles influence the amount and distribution of the solar energy that irradiates the Earth's surface.

In addition, it has been recently shown that the impact of such changes on the global climate is amplified by feedback effects. A first example is the albedo, connected to the reflection of sunlight from the Earth's surface, which increases in proportion to the size of glaciers and contributes to the decrease in temperature. Another example is the outgassing of CO_2 (the most abundant greenhouse gas) from the oceans, which increases during the hottest periods, resulting in a further increase in global temperature.

The validity of the astronomical model proposed by Milankovitch to explain the climatic fluctuations of the Pleistocene was confirmed in the 1950s and 1960s by the study of oceanic sediments. Here, the environmental information is stored in the remains of tiny single-celled marine organisms, the foraminifera. These protists, whose size can reach to about 10 millimeters, derive their name from the myriad of small holes that cover their calcareous exoskeletons. Foraminifera are of two kinds: the benthic type, which lives in deeper waters, and the planktonic type, which lives closer to the surface. Each type also comprises many species adapted to different water temperatures: the distribution of the different species of foraminifera, therefore, provide clues to past climate. Specific environmental parameters are connected to their intimate structure, that is, to the oxygen isotopes that make up the calcium carbonate ($CaCO_3$) of which their exoskeletons are made.

In the early 1950s, American nuclear chemist Harold Urey noticed that there was a relationship between sea surface temperature and the relative concentrations of different oxygen isotopes in the shells of shellfish (oxygen includes three stable isotopes: oxygen-16, oxygen-17, and oxygen-18, the first being the most common). Urey quickly realized that this effect could be exploited to develop a "paleo-thermometer" capable of providing temperature variations during the deep past. But it was Urey's student, the Italian Cesare Emiliani (1922–1995), who applied the method to the foraminifera, ubiquitous in oceanic sediments. The application of this method showed that temperatures had changed considerably over the last few hundred thousand years. Emiliani developed a system to classify climate periods according to different temperature values obtained through the measurement of oxygen isotopes.

These periods, connected to cyclic climate fluctuations, are called oxygen isotope stages (OIS) or marine isotope stages (MIS). By convention, stages designated with an odd number refer to warm periods, during which the relative concentration of oxygen-18 was relatively low. The current interglacial period, the Holocene, is therefore defined MIS 1. The next odd numbers refer to previous interglacial periods. The even-numbered isotope stages define cold periods, which are characterized by an enrichment of oxygen-18 in the minerals precipitated from seawater.[3] MIS 2 is thus the Last Glacial Maximum, whereas the subsequent even numbers refer to the previous ice ages. The appearance of *H. sapiens* in Africa took place at the beginning of the glacial stage MIS 6, a period of harsh and variable climatic conditions. During the same period, or probably shortly before, Eurasia was occupied by *H. neanderthalensis*, a species that was well adapted to cold climate and that was able to survive through several glacial and interglacial stages. The common ancestor of *H. sapiens* and *H. neanderthalensis* probably came from Africa during the Middle Pleistocene, when global climate changes, including surges and drops in temperature, were much more intense.

Intermediate events at a finer resolution are called stadial (cold phases during an interglacial period) and interstadial (warm phases during a glacial period). The use of odd and even numbers follows the same convention described above. As an example, the last interstadial is MIS3, between 60 and 25 KYA, which was discussed in previous chapters. It is the period in which *H. sapiens* dispersed, first from Africa to Asia and Australia and then to Europe, with possible impacts on the extinction of animals (such as the Australian megafauna) and of other human species (such as Neanderthals).

The analysis of oxygen isotopes in foraminifera from ocean sediments confirmed Milankovitch's theory: astronomical cycles mark global climate changes during the Pleistocene, as reflected in the average temperatures of the ocean surface and, as we shall see later, in the sea level and the volume of glaciers. In different parts of the world, similar cycles characterize rainfall, drought, and other climate indicators. Global climate dynamics, however, are complicated by atmospheric and oceanic circulation and by various feedback effects.

During the last two decades, ice cores extracted from glaciers in Greenland and Antarctica have also provided a reliable record of past temperatures with high temporal resolution. The environmental history of the last 800 thousand years has been reconstructed using ice cores more than 3 kilometers long. The ice is formed through the compaction of the snow that falls after the evaporation of sea water. Molecules containing the lighter isotope, oxygen-16, have a greater chance of evaporating compared with the heavier ones containing oxygen-18. At the same time, heavy water molecules condense more readily. Hence, the concentration of oxygen-16 in precipitation at the polar regions is higher when the global climate is warmer (causing an increase of heavy oxygen in ocean waters). During the same periods, the lightest isotope of hydrogen, hydrogen-1, is also enriched in the ice with respect to hydrogen-2 (deuterium). As a result, the concentration of oxygen and hydrogen isotopes in ice cores allows us to reconstruct global temperature fluctuations in the deep past. Between 2.6 and 1.1 MYA, a full cycle of glacier advance and retreat lasted 41 thousand years, which corresponds to the obliquity period. On the other hand, the cycles of the last million years had a period of 100 thousand years, compatible with the time span of eccentricity variation. During the last 2 million years, the glacial phases lasted on average 26 thousand years, whereas warmer interglacials periods lasted 27 thousand years. The recent analysis of deuterium in Antarctic ice cores, as part of the European Project for Ice Coring in Antarctica (EPICA), confirmed that the glacial cycles of the last 800 thousand years were determined by the combined effects of obliquity and precession of the equinoxes, whereas the cycle of precession variations prevailed only in the last 400 thousand years, with a period of 23 thousand years.

As we have seen, there is a continuous improvement in the quality of proxy data on past climate variability deduced from ice cores, sediments, stalagmites, and other natural archives. These data, characterized by increasing temporal and spatial resolution, provide a reliable input for the definition of models of paleoclimate variability.

Such global climate models are based on equations that describe the fluid dynamic processes and radiative/convective energy transfers that determine Earth's climate, including feedback effects such as albedo, interactions between atmosphere and biosphere through the carbon cycle, ocean evaporation, and changes in ocean currents.

In addition, regional climate models can be used to assess the distribution of precipitation and temperature in specific areas of the world during periods of interest for the study of human evolution and dispersals. Models can include the impact of climate change on vegetation, fire regimes, the hydrology of lakes and rivers, and the dynamics of glaciers. The correct interpretation of the information stored in the geologic record also requires an understanding of the interaction between animals and plants, the biogeochemistry of lakes and oceans, and the isotopic fractionation induced by precipitation and evaporation.

Tracing Past Environmental Change

The pollen of plants that flowered in the deep past, carried by the wind, often ended up in the sediments of surrounding lakes, seas, or wetlands. This information can now be recovered, as we have seen, by extracting cores that preserve the history of the geologic deposit and provide detailed information on the vegetation that existed in different periods of the past. In fact, pollen is preserved for millions of years thanks to the outer wall of sporopollenin, one of the most durable biological materials that exist in nature. A microscopic analysis reveals the male germ cells of the plants, with diameters ranging from 10 to 100 microns, the structure of this material, and the different fascinating forms of the granules. Furthermore, each species of plant has its own typical shape of pollen: triangular for eucalyptus, similar to a soccer ball for wattle, and so on.

Let's consider, for example, the pollen found in the sediments of Lynch's Crater, a volcanic lake in northern Australia. The lake formed more than 200 KYA, when the volcanic activity ended, and it provides today one of the longest archives on the environmental history of that

Australian region, including the effects of the last two glaciations. The lacustrine sediments contain not only pollen but also, among other materials, carbon particles. These may be a signature of vegetation fires, critical events in the history of humans and other animals. The archives of Lynch's Crater allow us to reconstruct the transformation of the Australian environment from rainforests into sclerophyll-type plants, the so-called bush, which consist of various species of eucalyptus trees that not only stand up well to fire but are also helped by it in their growth and diffusion.

The pollen shows that in the period corresponding to the oldest part of the sediments, the lake of Lynch's Crater was surrounded by large *Araucaria* conifers, which were growing all over the planet until the Cretaceous—and then, since the beginning of the Cenozoic, were able to survive only in the southern hemisphere. There were also forests of *Podocarpus*, typical conifers of the southern hemisphere that had evolved on the Gondwana continent but were very susceptible to fire. By using radiocarbon dating on small (weighing less than one milligram) fragments of coal found in the sediments, researchers from the Australian National University identified an increase in the frequency of fires around 42 KYA. The new landscape was dominated by fire-tolerant sclerophyll bush, such as eucalyptus and acacia trees. These species prevailed until the end of the Pleistocene, when a more humid climate favored the expansion of other plants, such as red cedar and laurel, which now mix with eucalyptus.

When the timing of the fires that suddenly burned vast regions of the continent was confirmed, Australians paleoecologists understood that in the sediments of Lynch's Crater they were reading the history of the first *H. sapiens* groups as they spread across the continent, burning woods to control hunting areas. This environmental impact would eventually contribute to the extinction of the Australian megafauna. But not everybody agrees on this interpretation, and the question remains open.

The sediments of another lake, this time in Italy, tell a similar story of climate and humans. The drama took place in a period close to that of the Australian event, but the victims were not gigantic hippos with camel-like muzzles, marsupial lions, lizards as large as tanks, or wingless birds weighing more than a ton.[4] This time, it was an entire human species that disappeared: *H. neanderthalensis*. Analysis of the pollen in the sediments of Lago Grande di Monticchio, in Lucania, Italy, shows that 40 KYA there was a sudden increase in aridity coinciding with the eruption of the Campanian ignimbrite (dated using argon-40/argon-39 to 39.5 KYA) at Campi Flegrei, north of Naples—one of the greatest volcanic eruptions in the Mediterranean over the last 200 thousand years. It has been estimated that two trillion tons of sulfur were emitted, a quantity comparable with the emission of Toba (the eruption that occurred in Sumatra 74 KYA). In fact, the layer of pumice and ash produced by the Campanian volcanoes is clearly visible in the sediments of many regions of the Mediterranean. The lacustrine cores show that the Mediterranean climate became quite variable throughout MIS 3, with warm and cold periods alternating over a relatively short time, even on a century scale. Isotopes in ice cores from Greenland show similar temperature changes, suggesting a global effect.

As a rule, when temperatures dropped, ice sheets expanded and forests were replaced by tundra and steppe. The opposite occurred during warmer periods. Such changes in vegetation were reflected in the types of animals available for hunting, making life difficult for the small groups of Neanderthal hunter-gatherers, who dispersed periodically to escape the extreme climatic conditions that were occurring cyclically. The presence of *H. sapiens,* who crossed the vast steppe that stretched from Central Asia to Italy's Po Valley to arrive in the Neanderthals' lands, did not make life easy for the human species that had dominated the Eurasian scene for hundreds of thousands of years. Around 30 KYA, or possibly earlier as we mentioned before,

H. neanderthalensis, already stressed by a period of rapid and intense environmental changes, went extinct.

Among the climatic conditions of this particular period of MIS 3, there is a short but intense cooling peak around 38 KYA, called Heinrich event 4 (H4). This is one of the six Heinrich events identified in the ice cores that record the last Ice Age, the last four of which occurred during MIS 3. These climatic episodes were characterized by sudden drops of 2°C–3°C in global temperatures, which lasted only a few decades. One hypothesis is that they occurred when huge icebergs broke off the North America Laurentide ice sheet, which had become unstable after growing to continental proportions. These blocks of ice moved across the North Atlantic, melting eventually. The subsequent injection of cold water of low salinity may have altered the "thermohaline circulation," the ocean currents that, as large conveyor belts, distribute the heat produced by solar irradiation, and which are controlled by the temperature and density of the water.

H4 coincides with a quick excursion of the geomagnetic field (the so-called Laschamp excursion). This minimum in the geomagnetic field produced a sudden increase in the concentration of long-lived cosmogenic radionuclides such as beryllium-10, measured in both lake and marine sediments and in cores from Greenland and Antarctica.[5] Geomagnetic and other natural markers (including ash layers and isotopic spikes) are useful to synchronize over extended regions the geologic clocks we use to assess the timing of climate change, the biological evolution of plants, animals, and humans, as well as the cultural evolution of our species.

Sea Level and Human Dispersals

Coral forms underwater, but when the sea level rises, for example due to the melting of polar ice, the increased depth of water absorbs the Sun's radiation, which cannot reach anymore the algae that live symbiotically with the coral. Through photosynthesis, these microscopic

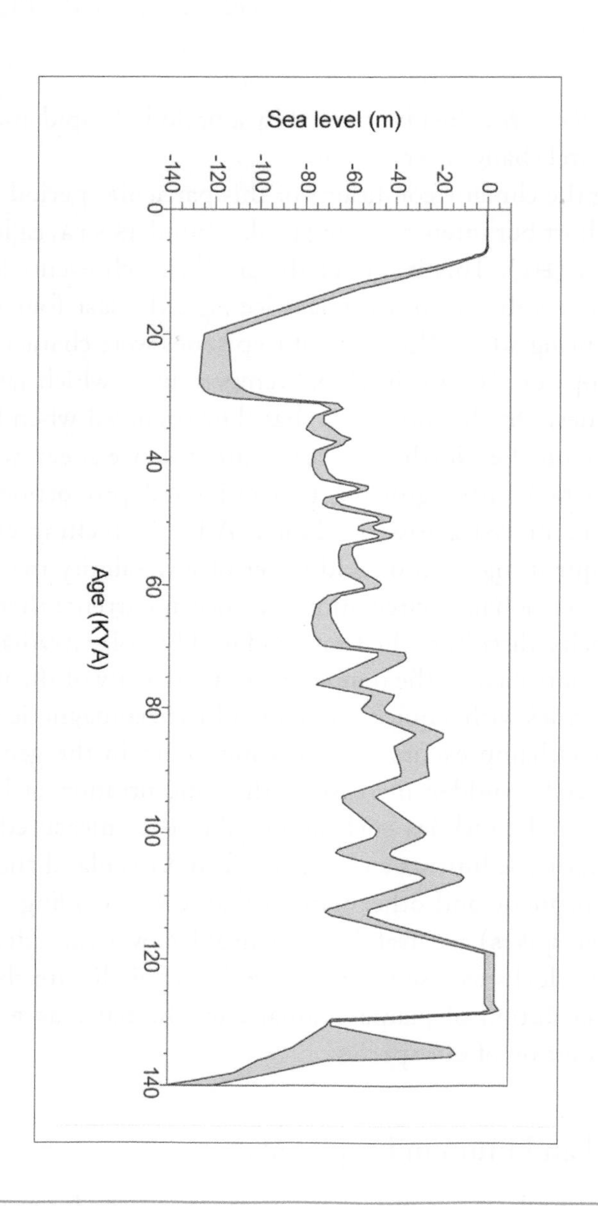

FIGURE 3.1
Sea level changes over the past 140 KYA. The chart is based on uranium-series dating of the Huon Peninsula coral terraces on the northern coast of Papua New Guinea. From Tuniz et al. (2009).

plants provide the corals with necessary nutrients, so the absence of solar energy kills the algae and, consequently, the coral reef.

The corals of the Huon Peninsula, on the north coast of Papua New Guinea, are an interesting case. Here the change in sea level is accompanied by the effects of plate tectonics, and the end result is a unique record of climate variability during the Pleistocene. The coast and the ocean margin in this area are situated where the Australian continental plate is headed on a collision course with the western Pacific plate. The phenomenon of subduction gradually raises the lithosphere, on which the coral terraces develop, with a speed of 4 millimeters each year. Upon the arrival of an interglacial period, the sea level also rises, but at a higher rate than the tectonic uplift, until it becomes stable at a new level and promotes the growth of a new terrace of coral. Hence, the different steps of the staircase that today rises for a kilometer into the sky on the Huon Peninsula correspond to the final parts of the glaciations that have taken place periodically over the last hundreds of thousands of years. The flat surface of each step represents the death of the anthozoans that built the coral, and it is followed (and preceded) by an inclined surface that corresponds to the establishment of a new generation of "animal flowers" on the skeletons of their predecessors. We can thus reconstruct past sea levels by precisely dating the coral steps. Of course, to calculate the level of the sea at the points of coral growth we must subtract the amount by which the sea level had risen as a result of tectonic effects.

Radiocarbon is not very precise in coral dating. When the aragonite produced by anthozoans partially turns into calcite,[6] there is a carbon isotope fractionation that can alter the real age by thousands of years, introducing large errors into the chronology. Hence it is preferable to use the uranium series method.

Natural uranium, which is present in seawater at concentrations of three parts per billion (or 3 milligrams per cubic meter), is incorporated into the limestone structure of the coral. Uranium-234, one of its radioactive isotopes, decays into thorium-230, which is also incorporated into the coral. The different chemical properties of uranium and thorium allow us to use the radioactivity of uranium-234 as a geological clock. In fact, whereas uranium is soluble in water, thorium is not, and hence it is not found in natural waters. Minerals precipitated from ocean waters do not contain primordial thorium, so all the thorium-230 measured in the coral was produced by the decay of uranium-234. The clock starts at the time when the anthozoans began precipitating the crystal. Uranium-234 decays into thorium-230 with a half-life of 245.5 thousand years, and the decay of thorium-230 into radium-226 has a $T_{1/2}$ of 75.38 thousand years. This series of decays reaches the so-called secular equilibrium after about 500 thousand years. At this time, the concentration of uranium-234 becomes the same as that of thorium-230 and remains constant; therefore, the clock no longer works. The maximum age that can be determined via this method depends on the sensitivity of equipment available to measure the ratio between the concentration of thorium-230 and uranium-234 close to the point of secular equilibrium, when this ratio remains relatively stable over time.

In the past, this type of analysis was performed by counting the alpha particles produced by radioactivity and allowed to measure a maximum age of about 250 thousand years. A more advanced method for this analysis is secondary ionization mass spectrometry (SIMS), which directly counts the atoms of uranium-234 and thorium-230 without waiting for their decay, making it possible to obtain ages up to 500 thousand years.

It was thus established that during the Last Glacial Maximum the sea level was 120 meters lower than in the current interglacial period. The dating of the coral stairs in Papua New Guinea also allows us to determine that during the interstadial period of the last Ice Age there were sharp rises in sea level. For example, 16, 22, 30, 38, 45, 52, and 65 KYA the surface of the oceans rose by 10–15 meters, probably due to the collapse of glaciers that caused the Heinrich events mentioned earlier. It was during this period of instability (MIS 3) that the major dispersals of our species occurred together with a significant behavioral evolution.

We already mentioned that one of the theories of human dispersal suggests that an originally small group of modern humans left the African continent around 60–70 KYA. Other theories postulate the existence of multiple human waves in earlier times. Generation after generation, always staying in the tropical area and near the coast, *H. sapiens* crossed the Arabian Peninsula and the Indian subcontinent, finally arriving in the islands of Southeast Asia, which during the ice ages were connected by land bridges all the way to the present island of Bali. According to many archaeologists and paleoclimatologists, the first dispersal of *H. sapiens* out of Africa occurred along tropical trajectories during alternating periods of marine transgressions and regressions. These events created an increase of coastal resources at lagoons and estuaries, encouraging early humans to develop navigation practices in order to exploit them. In contrast, sea level increases during Heinrich events flooded coastal areas and small islands, forcing the dispersal of animals, including humans.

Climate and Extinctions

During MIS 3 and MIS 2, but also in later periods, various species of animals, including some human species, went extinct. This happened in conjunction with the arrival of modern humans into different regions of the world and sometimes coincided with extreme climatic events.

At the time of H5, during the warm period of MIS 3 (about 45 KYA), fifty animal species disappeared in Australia, representing 90 percent of land animals weighing more than 50 kilograms, including the marsupial lion *Thylacoleo carnifex*, the lizard *Megalania prisca*, and the giant bird *Genyornis newtoni*. These extinctions have been studied since the early 1800s, when paleontologists first described the massive bones of ancient animals that had been discovered in caves and lake basins by the early Australian explorers. Richard Owen, the English anatomist who was the first director of the Natural History Museum in London, considered the "hostile agency of man" as a possible explanation for the extinction of these large Pleistocene animals.

Today, many scientists believe that the megafauna extinction was due to the combined effect of climate and human impacts. At the time, Australia was characterized by climatic and environmental conditions that were less harsh than in previous periods. According to many scientists, this evidence can be used to blame humans as the main cause of the megafauna extinction in Australia. Other scientists believe, however, that climate variability during the previous 100 thousand years or more had caused environmental stresses that large animals were unable to cope with, bringing their numbers close to a critical threshold for the survival of the species. Hence, human intervention would have been just the coup de grâce.

To resolve this issue, several researchers are attempting to improve the chronology of Australian archaeological sites through remains of humans or megafauna (sites in which both are present, such as Cuddie Springs, are rare). The dates available point to a very short overlap between humans and megafauna: not more than 3,500 years, around 46–47 KYA. This suggests that the human impact might have been the cause of the megafauna extinction, even though it is quite reasonable to assume that climate change also played its part. The debate on the respective role of humans and climate change in the extinction of Australia's megafauna during the Pleistocene is thus still very much alive. On this theme, opposing factions are engaged in a fierce battle, also with political implications (such as the rights of current Australian Aborigines to manage national parks). Much work remains to be done in different areas, from archaeology and paleoanthropology to radioisotope geochronology.

A similar story surrounds the extinction of the Neanderthals. As we have seen, about 30 KYA or earlier, during the H3 event, Neanderthals, probably already stressed by previous climatic events such as H4 and H5, became extinct in Eurasia. The combination of climate variability and the arrival of *H. sapiens*, the great predator anatomically and culturally "modern," was fatal to them. More or less during the same period, almost 40 percent of large animals in Eurasia become extinct, including cave bears, mammoths, hyenas, and other species that had survived for a relatively long period after the arrival of *H. sapiens*. (However, mammoths survived until 4 KYA on the island of Wrangel, northern Siberia.)

During this final phase of the Pleistocene, other enigmatic humans, different from both *H. neanderthalensis* and *H. sapiens*, disappeared: *the Denisovans*. So far, only scarce remains of this species (a phalanx of the little finger, another one of a toe, and two teeth) have been found in Denisova, Siberia. The same fate was met by the controversial *H. floresiensis*, found in 2003 in the Liang Bua cave on the Flores island in Indonesia. It is not clear when *H. erectus* disappeared; until recently he was considered to be still present in Southeast Asia (Solo River sites) at the time of the fifth Heinrich event, 45 KYA, but the new dates for these sites (using argon-40/argon-39, ESR, and uranium-series) are older than 500 KYA.

Conventional wisdom (which is being challenged by recent theories) holds that *H. sapiens* populations dispersed from eastern Asia to North America—where their lithic industries have been found in the archaeological record—around 13 KYA, in coincidence with a sudden cooling event that affected the northern hemisphere, the Younger Dryas. The origin of the Younger Dryas is similar to that of the Heinrich events we discussed earlier. In this case, too, the melting of ice in the northern hemisphere injected a huge mass of cold water with low salt content into the Atlantic Ocean, disrupting the ocean currents that carried heat from the southern to the northern hemisphere.

It was American scientist Paul Martin who, in 1984, first hypothesized that humans had played a role in the extinction of the American megafauna. The arrival of *H. sapiens* on the American continent coincides with the extinction of 135 animal species, including terrestrial giant sloths, saber-toothed tigers, lions, cheetahs, mastodons, mammoths, camels, and several species of horses and bison. These animals had survived many previous climate fluctuations, and the origin of some of these species went back to the beginning of the Quaternary, when North and South America joined together. The climate change affecting the final stage of the Pleistocene was less severe than during previous periods—with the exception of the changes induced by the already-mentioned Younger Dryas. The bones of the American megafauna are recent enough that they can be dated with radiocarbon. Unfortunately, radiocarbon dates for this period carry greater uncertainty due to the calibration curve, characterized by strong fluctuations of the radiocarbon concentration in the atmosphere. In any case, it is not yet proven that the Younger Dryas or other climatic events had such an impact on the American megafauna species as to cause their extinction. Although climate probably contributed to their demise, the admonition of Paul Martin still remains true: "The arrival in new lands of modern humans, with sustainable population growth, announces a bad omen for all animals with dimensions similar to or larger than that of the humans."

Other extinctions hit small islands, again in coincidence with the arrival of humans. In Madagascar, where *H. sapiens* first arrived only about 2 KYA, the large bird *Aepyornis maximus* and various species of lemurs and other animals disappeared. The same happened in New Zealand, where the arrival of modern humans caused the extinction of the moa,[7] and in the Caribbean islands, where terrestrial sloths went extinct 5 KYA.

FIGURE 3.2
(*Opposite*) Correlation between stratigraphic, paleomagnetic, and paleoclimatic records during the Quaternary.

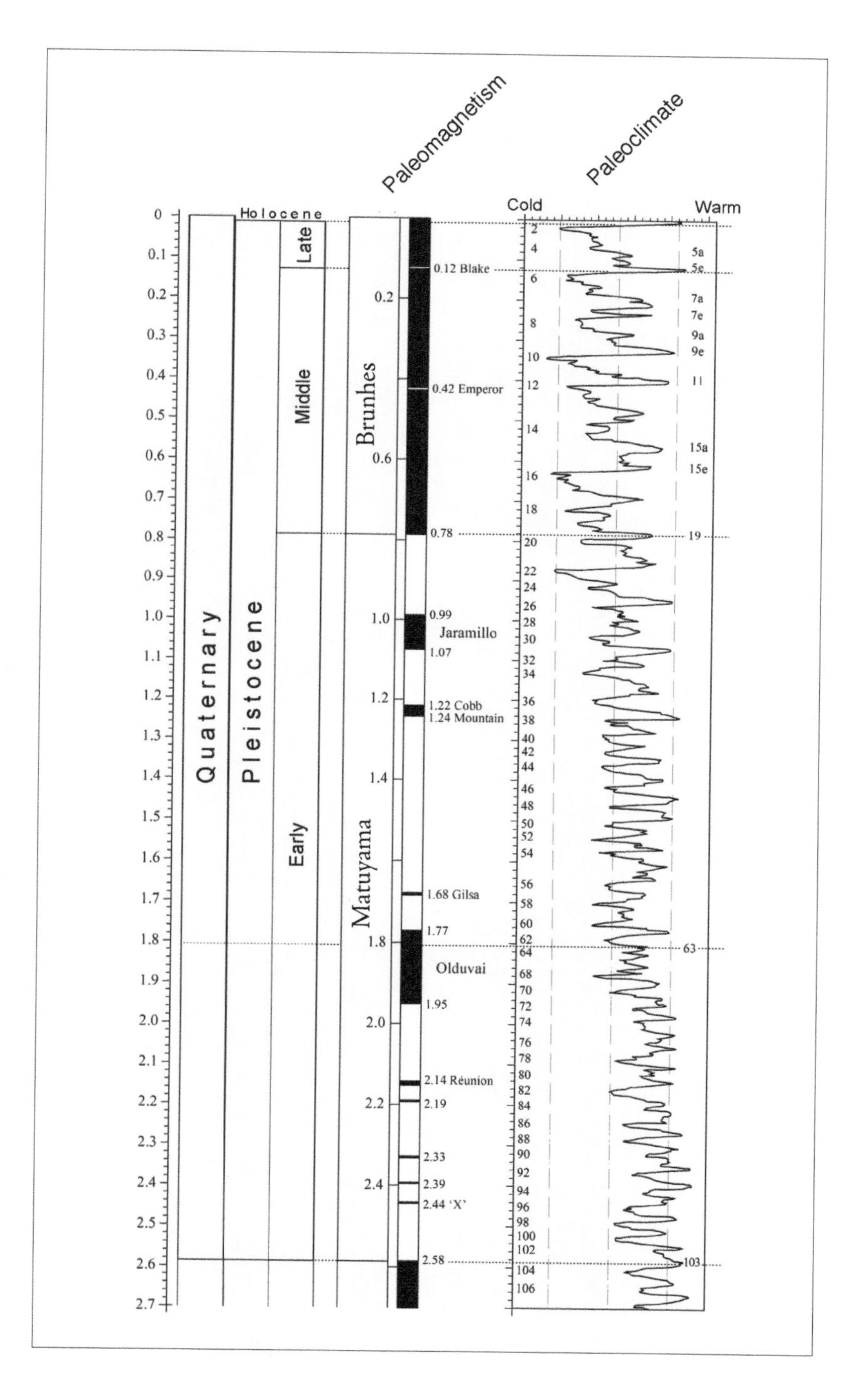
Paleomagnetism
Paleoclimate
Cold
Warm
Holocene
Late
Middle
Early
Quaternary
Pleistocene
Brunhes
Matuyama
0.12 Blake
0.42 Emperor
0.78
0.99
Jaramillo
1.07
1.22 Cobb
1.24 Mountain
1.68 Gilsa
1.77
Olduvai
1.95
2.14 Réunion
2.19
2.33
2.39
2.44 'X'
2.58
5a
5e
7a
7e
9a
9e
11
15a
15e
19
63
103

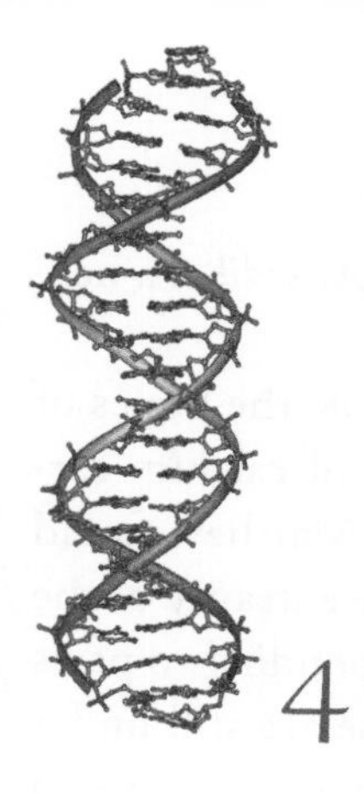

4

New Microscopes and Quantitative Paleontology

Virtual Paleoanthropology

With surgical precision, first he cleaned the cranium from the crust covering the outer surface, its base, the orbits, and other cavities. . . . Then he started slicing the braincase, revealing all the details of its internal structure, which reflected the shape of the brain surface. He found that the endocast had a volume of just over 400 cubic centimeters, almost like that of a chimpanzee, but the frontal lobes showed asymmetries similar to those that characterize the human brain. . . . Eventually he focused his attention on the teeth. Thus, he slowly pierced the layers of enamel down to the dentine and pulp chamber, noticing that the growth lines revealed a pattern of development halfway between that of a chimpanzee and a human.

If you are thinking that these operations were performed manually, with the use of a scalpel or some other surgical instrument, you are offtrack. Indeed, this entire procedure was carried out in a virtual way, with full approval of the staff of the Museum of Moropeng,

South Africa—the curators of the precious skull of *A. sediba,* a new species of *Australopithecus.*

For almost two million years, until August 15, 2008, the bones of *A. sediba* had been preserved cemented in a breccia of calcium carbonate. On that August day, Lee Berger and his son Matthew found some parts of the skull (a jawbone and a tooth) at the entrance of the Malapa karstic cave site. Other fossil remains corresponding to parts of different skeletons were discovered later, and these are still under excavation and study. It wasn't easy to find them, as they were covered with a crust of limestone that camouflaged the fossil bones among the stones at the entrance of the cave. For years the paleoanthropologist from the University of Witwatersrand had been looking for possible sites of our ancestors, exploring this area of South Africa using also Google Earth (after the correction of GPS data, which are altered for military security reasons). Malapa is part of an area that has gained UNESCO World Heritage status by virtue of the large number of

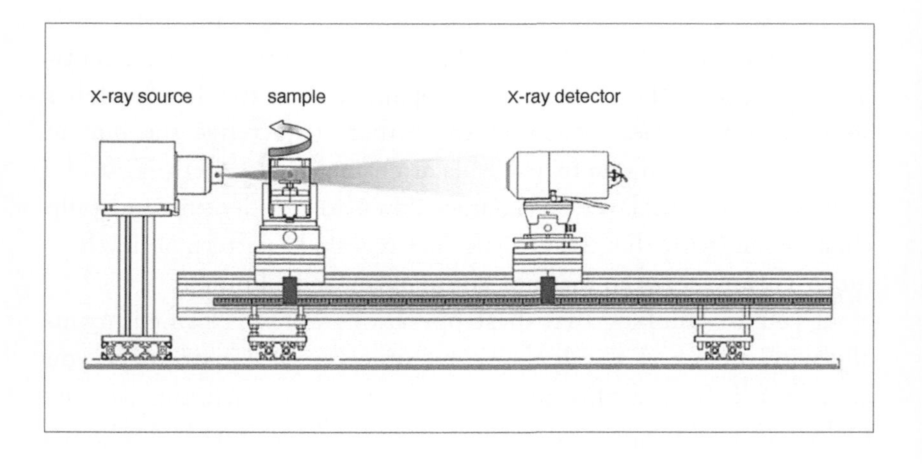

FIGURE 4.1
Schematic drawing of a portable microCT system.

hominin finds, more than five hundred. The first of these was the skull of the so-called Taung child, type specimen of another species of *Australopithecus*: the *A. africanus*, discovered in 1924 and described the following year on the journal *Nature* by the anatomist and paleoanthropologist Raymond Dart.

Thanks to new analytical tools, today precious remains like the skull of *A. sediba* can be studied in detail without invasive interventions. During the last twenty years, virtual paleoanthropology has become a reality, following the revolution in X-ray imaging. The new microscopes use powerful particle accelerators, such as the synchrotron light source of the European Synchrotron Radiation Facility in Grenoble, France, or Elettra, the synchrotron accelerator in Trieste, Italy.

The methodology used in the operations described above is based on computed microtomography (microCT), a method that allows one to slice (virtually) the precious fossils, showing their internal details with high precision. The high energy X-rays can easily penetrate a fossilized skull, which has the same density as rock. The virtual blades that cut the bone are powerful mathematical algorithms that, thanks to the continuing progress in the field of computer microprocessors, are able to operate rapidly on the enormous number of voxels— terabytes of data—in which the fossil is virtually disintegrated.

"Light will be thrown on the origin of man and his history," alluded Charles Darwin at the end of *On the Origin of Species*, unaware that a new light (the synchrotron light) would become available over a century and a half later to reveal our origins.

Brief History of Paleoradiology

"I have seen my own death," said Anna-Bertha Röntgen in 1895 when her husband Wilhelm obtained the radiography of her hand, complete with wedding ring. The image was produced on a photographic plate by the mysterious radiation that the German physicist Wilhelm

Conrad Röntgen had discovered. Using a simple piece of paper covered with platinum barium cyanide, he had noticed that some luminosity was produced by mysterious rays (which he called X-rays) emitted while he was studying the effects of electrical discharge through a gas at low pressure. This occurred even when the fluorescent sheet was shielded or located in a different room. It was soon realized that X-rays could have important applications, especially as a diagnostic tool in medicine. They allowed one to see not only the structure of the bones but also the internal body organs. Marie Curie promoted their use in the battlefields of World War I to locate bullets and bomb fragments in the bodies of the soldiers.

The same year of their discovery, X-rays were used to "see" inside Egyptian mummies and check their state of preservation and other details without removing their bandages. They were applied soon also in paleoanthropology. In 1903, only eight years after Röntgen's discovery, the German dentist Otto Walkhoff took radiographies of the Neanderthal mandible discovered at La Naulette (Belgium) in 1866. In 1906, the Croatian paleontologist Dragutin Gorianović-Kramberger used X-rays to study the bones of the Neanderthals from Krapina that had been discovered in 1899. He was able to study the morphological details of the fossil teeth of *H. neanderthalensis*, discovering for example that our cousins were affected by taurodontism.

Over time, X-ray detection techniques became increasingly refined. The first improvements concerned the phosphorescent materials and photographic films used for the production of radiographic images. Then, physicists developed special scintillating materials that could produce light when bombarded by X-rays. This light was in turn converted into electrical signals with very high spatial resolution. The latter conversion was based on the "photoelectric effect," the emission of electrons by atoms bombarded with photons.

In the 1970s, the advent of digital technologies and the spread of increasingly powerful computers, both in terms of processor speed and memory capacity, revolutionized X-ray imaging. A special invention of this period was computerized axial tomography (we will use the term "computerized tomography," or CT), developed in 1972 by Godfrey Hounsfield, who, along with South African physicist Allan Cormack, received the Nobel Prize in Medicine in 1979 for this discovery. CT systems are based on the use of multiple radiographies taken at different angles around a rotation axis. These 3D images, including inner sections of the analyzed object, are then reconstructed from X-ray projections using mathematical algorithms developed in 1917 by Austrian mathematician Johann Radon (obtaining the so-called back projections). Hospital CTs, in which the X-ray generator and detection system rotate around the body part to be analyzed, achieve spatial resolutions of around half a millimeter (0.3 millimeter in the most advanced models).

X-ray CT analyses were immediately applied to various archaeological and paleoanthropological cases. For example, the CT scan of the Taung child, performed in 1984, showed that its dental development had similarities with that of apes. In 1994, CT analyses were performed on the inner-ear channels of *Australopithecus* and *Homo* to show that they were related to their different types of bipedalism. This study was then extended to the inner ear of the Neanderthals. Even Ötzi, the mummy from Tyrol we mentioned above, underwent X-ray CT analysis. Its skull was virtually reconstructed and its body was analyzed in detail to try to understand whether some deformations were due to ice pressure or wounds from a weapon. The analyses indicated that the Similaun Man had bone fractures to both its nose and its ribs. It also had an arrowhead injury in the shoulder in correspondence with the position of an artery. This was probably the cause of the man's sudden death.

Since then, many other mummies found in Peru, Egypt, Siberia, China, and other localities have been virtually "stripped" by X-rays to expose their inner details.

A Powerful Light

We already mentioned that a circular accelerator (the synchrotron) can produce an intense X-ray flux, which makes possible CT imaging with submicron spatial resolution (microCT).[1] As in conventional CT, microCT images can be displayed in various ways using advanced software programs. We can obtain sections of an object in all the desired orientations or show its external and internal 3D structures. We can, for example, create the image of the inner surface of a skull and then reconstruct the surface of the brain, as mentioned above. We can also reconstruct the labyrinth of the inner ear, virtually remove single teeth from the jawbone, or separate the enamel from the dentine in a tooth. Finally, microCT images can be transformed into real three-dimensional objects using a procedure of 3D reproduction called

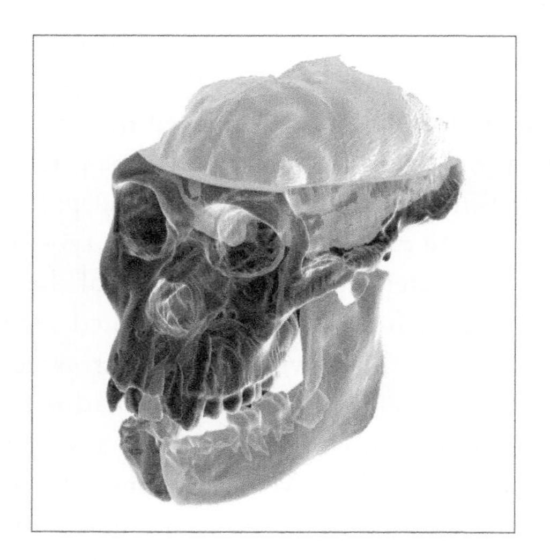

FIGURE 4.2
Reconstruction of the cranium MH1 of *A. sediba* based on digital data obtained with microCT. The external morphology of the brain is evidenced as well as the roots of some teeth and the crown of others that are erupting. Created by Kris Carlson, reproduced with kind permission of Lee R. Berger (University of the Witwatersrand).

stereolithography. MicroCT is therefore an advanced 3D microscopy that allows us to generate images of fossil teeth and bones with resolutions hundreds of times better than medical CTs. Obviously, unlike the latter, the sample irradiated with synchrotron light is rotated in front of the detector.

High-resolution tomography does not always need powerful accelerators. In recent years, microCT systems have been developed using microfocus[2] X-ray tubes and advanced detectors. Powerful portable devices, characterized by high spatial resolutions and efficiency,[3] can be operated directly in the museum in which the precious paleoanthropological remains are stored.

Let's now discuss, in more detail, the different applications of these new microscopes in studies of human origins.

Describing or Measuring?

In both their external and internal structure, the different parts of the human skeleton preserve an archive of the specific functions and thus the biological history of the organism. The skull is the most critical part, carrying information on many aspects related to human evolution, from cognition and communication to nutrition or even posture and the habitual locomotion.

A microCT of the braincase provides quantitative details on the brain structure, including the convolutions on its surface or the size and shape of its lobes. For example, the microCT of *A. sediba*'s skull reveals that its brain's right frontal lobe was larger than the left one, with a bulge in the inferior frontal gyrus on the left side. This structure does not exist in chimpanzees, whereas in humans it is related to the Broca's area, one of the main centers of language. These results suggest that the neural reorganization of the brain in our evolutionary line began before the volumetric expansion of the brain itself, with a possible increase of connections in the region that would become crucial for the development of our complex language and, therefore, our peculiar form of sociality.

X-ray microCT imaging can also be used to separate virtually the bone from the mineral matrix in which it was found. It is also possible to rebuild fossil skulls, which are often found in fragments. For example, the skull of *Sahelanthropus tchadensis*, which was discovered crushed and distorted in the desert of Chad, was difficult to characterize. Finally, its virtual reconstruction using conventional X-ray CT scanning suggested that it belonged to a bipedal primate, although this conclusion was controversial. To settle the issue, a high-resolution microCT of this skull has been performed at the European Synchrotron Radiation Facility in Grenoble, but the results have not yet been published. The Ceprano skull, which was found in Italy in 1994 and was later described as an ancestral form that preceded the divergence between Neanderthals and modern humans, at the time of discovery was partially deformed by the pressure of the sediments. MicroCT analyses are in progress to reconstruct its original conformation.

MicroCT analyses may be useful to identify the species to which a fossil belongs on the basis of just some bone fragments. For example, in 1995, a hospital CT was performed on a fragment of a fossil femur of unknown chronology discovered in Berg Aukas, in Namibia, in order to identify its origins. The microstructure of the cortical bone definitely related the bone to *H. heidelbergensis* and allowed the exclusion of other hypotheses (*Australopithecus*, "early *Homo*," *H. erectus*, and *H. sapiens*) that were consistent with other elements of the external morphology. Remains of fossil teeth have recently been used to discriminate between hominin and ancient ape genera using non-destructive microCT analyses of inner structures including enamel, dentine, the pulp chamber, and the enamel/dentine interface. In addition, these quantitative data can be used as input in comparative studies using statistics and cladistic analyses.

Over the past two decades, the development of these powerful new microscopes has been complemented by equally sophisticated innovations in the fields of morphology and morphometrics (the description and the measurement of biological forms, respectively).

An issue debated for centuries by morphologists (not just by paleoanthropologists) was the difficulty of reconciling the description of fossil findings with their measurement. In other words, since the beginning of paleontology, biological forms (e.g., a human skull, a femur, a tooth, or part of them) were at first described, with evident limitations due to the subjectivity of the evaluations made by each researcher and the inaccuracies of language with which these assessments were expressed. Such descriptions could provide a more or less accurate idea of the forms under examination, but they did not provide data that quantitatively expressed their shape and variability. By contrast, quantitative measurements of biological forms could provide numerical data of increasing accuracy. These measurements consisted mostly of distances between two points: for example, the length, width, or height of a cranium. Different angles and a large number of morphometric indices were also considered. However, it was difficult to integrate all these data.

This became possible in the second half of the twentieth century with the advent of computers, which made synthetic numerical evaluation possible by combining many of these measures through multivariate statistics. Important applications of multivariate statistics in morphometrics started only in the 1970s. Nevertheless, the quantitative data of traditional morphometrics were never combined with the subjective information of morphological evaluations.

The complexity of biological forms therefore remained inexpressible except by using a sort of double track consisting of synthetic but inaccurate descriptions on the one hand and measurements that were too limited and partial on the other. The real progress in the approach to the measurement of biological forms occurred in the last two decades of the last century (to become fairly widespread only after 2000) with the development of the so-called geometric morphometrics. This technique has been defined, with non-excessive emphasis, as the "new synthesis."

The new approach, also made possible by the use of computers of increasing power, has abandoned the diameters, angles, and indices of traditional morphometrics in favor of spatial coordinates (in two or three dimensions) of well-chosen homologous "landmarks," taken together as descriptors of the biological forms and their complex geometries. At this point, the potential of applying multivariate statistical techniques to this new database did the rest.

The new analytical procedures and related software have freed morphologists from the limitations of single measurements, allowing them to move toward a quantitative evaluation of biological forms in their entirety. Finally, then, the old morphological descriptions became measurable quantities that were reproducible and statistically comparable: the old dream of the morphologists finally came true!

The Obstetric Dilemma

Bipedalism, which was acquired during the early stages of human evolution, resulted in an increased risk at birth for both mother and child. This is because of structural changes in the mother's pelvis that forced the infant to perform complex twists during delivery. In addition, during the evolution of the genus *Homo*, the size of the brain, and hence that of the skull that must pass through the birth canal between the pelvic bones, increased more and more until it trebled!

The structure of the pelvis plays a crucial role in walking. The reconstruction—based on CT scanning—of the hipbones of "Ardi," the *Ardipithecus* female we met earlier, suggests an incipient bipedalism. At the same time, other elements of the skeleton indicate a highly flexible behavior in which the new bipedal anatomical features in the pelvis were combined with long arms and fingers, plus opposable big toes. All these functional features reveal that Ardi was still using trees, at least as a night shelter. One could think that, going back in time toward the common ancestor of chimpanzees and humans, the locomotion of our ancestors should be more similar to that of present apes. On the contrary, from the study of Ardi we now know that the locomotion models of the apes are probably different from those of our common ancestor. The common ancestor (to whom *Ardipithecus* must have been very close) was probably a "generalist," mostly arboreal. He was not, in any case, a knuckle-walker. However, the transformation of the pelvis into a shape compatible with bipedalism was still in its initial stage and thus did not create problems at birth (also considering that the *Ardipithecus* brain was still relatively small).

Another important female specimen that chronologically follows Ardi is "Lucy" (*A. afarensis*), whose skeletal remains are exceptionally complete. Its pelvic bone was wider and less flat than a chimpanzee's. The blade of Lucy's ilium (the uppermost and largest bone of the pelvis) was similar to ours: curved and concave, with part of the gluteal muscles attached to the lateral part of the pelvis, which acted on the femur to balance the posture. This allowed the australopithecines to walk without losing balance, still retaining the ability to conduct an arboreal life. In any case, the skull of the infant of *Australopithecus* was small enough not to create problems at birth.

Childbirth-related problems arose later, caused by the encephalization process during the evolution of the genus *Homo*. To understand the problems that bipedalism brings to this crucial life function, it is useful to remember the different birth modalities in chimpanzees and humans. With a brain volume smaller than that of the pelvic cavity, the baby chimp can pass directly through the birth canal, which has a constant cross section, and finally exit with its head back and its eyes turned toward the mother. These dynamics allow for a relatively easy childbirth. The process is much more complicated for the human child, with a massive skull that must pass through a birth canal whose major axis at the exit is perpendicular to the direction it has at the beginning, a structure produced by the evolutionary processes that led to the acquisition of bipedalism. The human child must first rotate to align the shoulders with the major axis at the entrance of the canal, and then turn again to align them with the major axis at its exit. At the end, human babies are delivered with the back of their heads turned toward the mother, after two half rotations. This makes childbirth in our species very difficult, increasing the risks for both the baby and the mother (in addition to causing an intense and prolonged pain), and usually requiring assistance.

In *Australopithecus,* the size of the head was smaller than the section of the pelvic cavity, as in chimpanzees, but a reconstruction of the different sections of the cavity suggests a composite process for the birth of these hominins: although they needed alignment at the entrance, as in *H. sapiens,* the little *Australopithecus* could then pass through the canal without further rotations, as in the case of chimpanzees. The female pelvis of the early *Homo,* and in particular of *H. ergaster,* was still fairly large and flat (platipelloid). It had probably evolved to allow the delivery of infants characterized by a larger brain, and it still entailed a mechanism of delivery without rotation, similar to that of *Australopithecus.* This hypothesis is confirmed by the discovery, in the area of Afar in Ethiopia, of a complete pelvis belonging to an adult *Homo* woman dating from about 1 MYA.

This type of pelvis is still found in archaic humans from the Middle Pleistocene, represented in Europe and Africa by *H. heidelbergensis,* the common ancestor between us and the Neanderthals, and similar characteristics are still found in these latter humans. Only with the emergence of our own species did the obstetric dilemma need to find an effective solution, with unexpected consequences, as we shall see.

It seems as though evolution was not able to continue changing the structure of the pelvis, as this skeletal component is critical for our balance in the upright posture and our bipedal locomotion. On the other hand, evolution could not intervene too drastically on sexual dimorphism, preventing females from walking efficiently in order to facilitate their reproductive vocation. It was possible, instead, to anticipate the timing of childbirth, and this is what happened. In fact, if you look at the primate family, *H. sapiens* shows two peculiarities: the abnormal size of the brain (more than three times that of a chimpanzee), and the relationship between brain size and gestation length. Let's briefly discuss this last point.

A typical non-human primate with a brain as large as ours should have (to be in line with other apes) a gestation period of eighteen months, whereas human pregnancies last only nine months. We also know that the first few months of our lives are characterized by the total lack of autonomy of the newborn, even in the few basic functions of survival. During this period, it is in fact the brain (and the nervous system in general) that completes its development, at a pace that we could call "fetal." Think of the fact that children, when they are still crawling, have an oversized head compared to the body: indeed, it's as if we completed our growth (especially that of our large brain) out of the womb, continuing our fetal life outside the uterus.

Now everything adds up: the eighteen-month gestation of the (human) primate with the large brain occurs half in the uterus and half outside. Otherwise our eighteen-month baby could not pass through the birth canal, so strongly modified by the new shape of the (maternal) pelvis evolved to accommodate bipedal evolution. This early birth had other interesting consequences, as we shall see immediately.

Teeth, Growth, and Development

The phenomenon of "early birth" was linked during human evolution to a more general trend toward a lengthening of the time needed to grow and develop. It is well known that our life history is longer than a chimpanzee's and that some of its main stages—such as childhood, adolescence, and the "post-reproductive" age—are especially longer. Only recently could these aspects of human and pre-human life histories be studied by using precious fossil teeth of extinct species and non-destructively examining their microstructure.

Of all tissues, teeth have the highest chance of remaining preserved in the fossil record. Since the early paleoanthropological studies, the analysis of their external morphology has provided valuable information on human evolution, but they can provide many additional details. In fact, the internal microarchitecture of teeth, such as the thickness and distribution of the enamel, as well as their growth structure and composition, preserves a detailed record on development, diet, migrations, pathologies, and stress: teeth are a real black box of our early life.

This information may be collected by sectioning the tooth and using conventional microscopy (confocal, polarized light, or scanning electron microscopy); but this would cause damage to rare and precious fossil remains. Furthermore, these techniques would only provide information on two-dimensional sections. Alternatively, X-ray microCT can reveal the tooth microstructure non-destructively and in three dimensions. X-ray absorption in the different structural elements of interest (enamel, dentine, roots, and jawbone) produces low-contrast images; hence, phase-contrast imaging is the preferred method for these kind of studies. The image resolution is comparable to that obtained through histological observations using a microscope, but samples don't need to be sectioned. This new approach could be called "virtual 3D histology." Since biological stresses are inscribed in the tooth enamel, it is even possible to identify with this method the trauma of birth.

Phase-contrast microCT has been performed at the Synchrotron Radiation Facility in Grenoble to study the teeth of the small *Australopithecus* from Taung. Counting the growth lines of the enamel, it is possible to calculate the body's age at the time of death and evaluate the length of the development age, which in general is shorter in apes and in pre-humans than in humans. The Retzius lines of this sample (measured with a spatial resolution of less than one micron[4]) reveal that the Taung child died at less than four years of age after having reached a level of development similar to that of a chimpanzee of the same age, whereas its dental age corresponded to that of a human child of five or six years.

Thus, the gap that we can observe between the parameters characterizing the life history of *H. sapiens* and that of the chimpanzee (our closest relative among the non-human living primates) probably did not develop during *Australopithecus* times, i.e., when the evolutionary scene was dominated by the acquisition of bipedalism. Instead, we can assume that these differences developed later, namely with the evolution of the genus *Homo*, when brain growth started driving the complex trajectories of human evolution. In fact, an increasing amount of data seems to confirm that the expansion of *Homo*'s brain was matched by a progressive lengthening of the growth and development periods, probably in parallel with the anticipation of birth that we mentioned in the previous section.

All this had other consequences—collateral effects, but with tremendous impact on our biocultural adaptation—such as the ability to learn. Prolonged childhood and youth are, in fact, consistent with

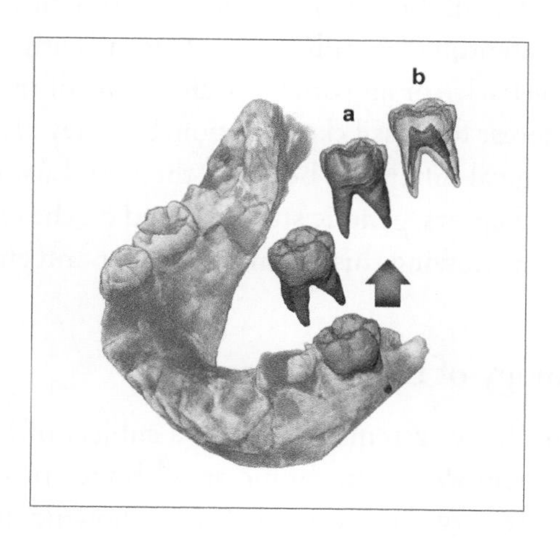

FIGURE 4.3
Juvenile Neanderthal mandible from Molare (Italy), with the lower-left second deciduous molar virtually extracted and dissected using X-ray microCT. The transparency of enamel shows the dentin (a) and the pulp chamber (b). From Tuniz et al. (2012).

a phenomenon that undoubtedly characterizes the evolution of the genus *Homo*: the production of stone artifacts of increasing complexity during the Paleolithic.

The biological archive preserved in fossil teeth tells us that even the Neanderthals had growth and development times different from those of *H. sapiens*, albeit in a modest and still not clearly quantified way; research thus continues with the new microscopes and modern morphological studies. These inquiries include the systematic analysis of enamel thickness using X-ray microCT and the measurement of the microspatial distribution of barium in enamel. The latter is a tracer of the mother's milk and can be used to compare weaning ages for Neanderthals and modern humans. The timing of such life history stages is of critical importance for its effects on population growth and species replacement.

If all the differences were to be conclusively demonstrated, we might well say that all the Neanderthal characteristics—including the bizarre skull, extending laterally and in length (with the well-known "chignon" in the back), or any other specific traits of the skeleton—are not only of interest for fossil classification, but they also represent the trace of a biological difference between them and us. As we shall see in the last two chapters, genetic studies based on the extraction of ancient DNA began showing this in a more proper and effective manner.

Otolaryngology of Extinct Species

Speech and auditory systems have been a subject of high interest for scholars of human evolution. Some small bones belonging to these systems can be preserved in the fossil record, allowing their direct study.

In particular, the hyoid bone plays a critical role in phonation and in the development of our articulated language. This small, horseshoe-shaped bone is located on the anterior midline of the neck, between the larynx and the mandible. Although it is not directly articulated with any other bone, the hyoid is connected through a range of muscles and ligaments to the tongue, the jaw, the larynx, the base of the skull, and the sternum.

Before discussing the study of the hyoid, let's summarize the main approaches followed to understand the origin of complex language. According to some views, language evolved to organize social activities that required coordination between several individuals, such as hunting large animals, competing with other human groups, or seafaring. Others believe that the evolution of phonation can be deduced from the morphology of the skull base—for example, from the size of the hypoglossal canal, which allows us to deduce the size and structure of the vocal tract, or from specific structures of the cerebral cortex related to language, such as the Broca area. Other opportunities are offered by paleogenetics, such as in the study of FOXP2, the gene related to linguistic capabilities, which will be discussed below. Finally, according to other perspectives, language could be related to the evidence of cultural developments such as the production of complex stone tools and the first manifestations of conscious thought and symbolism.

Although the mechanisms and structures used for voice production are similar in humans and other primates, our vocal tract is characterized by various morphological differences, including the structure and microstructure of the hyoid. Its position in the vocal system was reconstructed in the early 1970s in the Neanderthal skeleton from La Chapelle aux Saints by estimating the structure of the muscles that were attached to the skull base and to the inner part of the mandibular symphysis. These preliminary reports determined that the hyoid, hence the larynx, was located in the upper part of the throat in the Neanderthals, much like a newborn modern human or a chimpanzee, thus reducing the resonant space available for phonation. Initially, computer simulations suggested that our extinct cousins were using fewer vowels than *H. sapiens*, with more communication difficulties, but other studies show they had the same number of vowels as modern humans. Although the nature and origin of our peculiar linguistic faculty remains still open and controversial, recent studies claim that modern language developed 500 KYA and was already used by the common ancestor of modern humans and Neanderthals. This implies a slow incremental process of both genetic and cultural evolution. Does our language bear traces of the language spoken by the Neanderthals (or other human species)?

In the last few years, hyoid bones from different species, including *A. afarensis, H. heidelbergensis,* and *H. neanderthalensis,* have been discovered. Their study could provide direct information on the evolution of complex language. The hyoid of the Neanderthal discovered in Kebara (Palestine) has an external shape and size similar to that of modern humans, and this has led to the hypothesis that Neanderthals were able to speak normally. The availability of virtual histology based on X-ray microCT now offers the opportunity to verify, noninvasively, possible differences in the hyoid microstructure. The bone microarchitecture should reflect the different types of biomechanics related to different phonation performances. With the use of X-ray 3D images, following engineering methodologies widely developed in the biomedical field, "finite element analysis" models can provide the mechanical properties of different types of hyoid bones.

On the other hand, the use of language requires that modulated sounds arrive at destination in the brain through a system that can detect and transmit them. The ear performs this function through its different parts: external, middle, and inner, the latter "interfaced" to the brain. The pinna, the outer part of the ear, collects and transmits sound waves into the middle ear, where, after the external auditory canal and the tympanic membrane, a chain of small bones (the malleus, incus, and stapes) transmits the vibrations of the membrane to the inner ear (cochlea). Of these parts of the human auditory system only the three middle-ear bones can be found in the fossil record, usually near the ear canal of the skull. It was found that the Neanderthals from Atapuerca had malleus, incus, and stapes identical to those of modern humans, suggesting that they perceived sounds like we do.

In a recent study, the middle-ear bones belonging to the old remains of *P. robustus* from Swartkrans in South Africa, dating to 2 MYA, and those of an *A. africanus* from Sterkfontain (also in South Africa) have been compared with similar bones from present-day humans and apes. It has been shown that the malleus of these bipedal apes was very similar to that of *Homo sapiens*, whereas the morphology of incus and stapes was similar to that of apes. This means that their auditory system performed similarly to ours, but only in some respects: *P. robustus* and *A. africanus* must have still been hard of hearing, in contrast with some derived species of *Homo* such as Neanderthal and their direct ancestors. This is another element to add to the list of human characteristics that appeared in the early stage of our evolutionary history, a few million years ago. To traits such as bipedalism (with its effects on the birth canal), reduction of the anterior teeth, and reorganization of the brain (as we have seen for *A. sediba*), we must now add the human-like malleus.

The mechanical vibrations of the middle ear pass through the fluid and membranes of the cochlea in the inner ear, where they are converted into nerve pulses, which are in turn transmitted to the brain. The inner ear includes also the labyrinth, which consists of three semicircular canals filled with fluid, oriented perpendicularly along the three space planes and lined with receptors (hair cells) which, through the movement of the fluid, allow us to have a correct perception of our position in space and are crucial for balance. These structures of the inner ear are often preserved in fossil skulls, but they can be observed in detail only using X-ray CT.

The semicircular canals of *H. erectus* from Java were already studied using a hospital CT system thirty years ago, but the quality of imaging did not allow detailed investigations. During the 1990s, the improvement of the spatial resolution of X-ray CT allowed paleontologist

Fred Spoor to carry out pioneering studies on the semicircular canals of various Neanderthal skulls, revealing some differences from those in modern humans. In recent years, about twenty Neanderthals have been analyzed with CT scanning. The results confirm that their labyrinth was different from that of *H. sapiens* in size, shape, and orientation. This is an interesting finding, given that the labyrinth of *H. erectus* and early modern humans was like our own.[5] How can we explain the exception represented by Neanderthals? According to some interpretations, the particular shape of the semicircular canals in Neanderthals can be correlated with the need to control the movement and the rotation of their elongated head, located on a short and stumpy neck.

Besides X-ray imaging, isotopic microanalysis also provides crucial information in human evolution studies: we will give some examples related to hominin diets.

Prehistoric Restaurant

At the beginning of the Quaternary, vast areas of Africa, especially those on the eastern side of the GRV, suffered enormous environmental changes, becoming increasingly arid. Animals that failed to adapt became extinct. Nevertheless, the selective forces of evolution led to the emergence of new species of animals (including hominins) such as *Paranthropus*, also known as "robust" australo-pithecines. They could be found in eastern Africa (and later in southern Africa), dating to about 2.5 MYA. A well-known example was nicknamed "Nutcracker" by Mary Leakey. Its teeth were flatter and stronger than Lucy's, with thick enamel; it had robust mandibles and a characteristic ridge on the top of the skull where powerful chewing muscles were anchored. Hence Nutcracker could expand its diet, originally based mainly on fruit and leaves (typical of apes), to include roots, tubers, and nuts. The analysis of stable carbon isotopes in its teeth suggests that Nutcracker's diet was dominated by plant food of low quality.

Plants discriminate carbon-13 from carbon-12 through a process called "fractionation," the natural equivalent of the uranium enrichment used in the construction of nuclear weapons. Given that plants prefer the lighter isotope, the sugars they produce have a lower ratio of carbon-13 to carbon-12 compared with atmospheric CO_2. However, the magnitude of these fractionation effects varies according to different photosynthesis mechanisms. Carbon isotopes pass through the food chain, and their ratios in the tooth enamel[6] can tell us which vegetables were part of the preferred diet of extinct species: shrubs and grass using the photosynthetic mechanism C3, or C4 plants more adapted to drought and better able to hold water. C3 plants discriminate the heavier isotopes of carbon more efficiently than C4 plants.

Another hominin similar to Nutcracker, *P. robustus*, appeared in the same period in South Africa. It also had powerful chewing muscles, anchored to a similar sagittal crest on the skull, and massive molars, adapted to a plant-based diet. But the analysis of the carbon-13 to carbon-12 ratio in its tooth enamel seems to confirm that unlike Nutcracker, *A. robustus* fed also on various insects, including termites. The microanalysis of carbon and oxygen isotopes was performed across a section of its teeth with laser ablation inductively coupled plasma mass spectrometry (LA-ICP-MS), which also shows that this hominin adapted its diet to seasonal variations.

LA-ICP-MS was recently used to analyze carbon isotopes in phytoliths extracted from the dental calculus of *A. sediba*. The results revealed that the diet of australopithecines was based on C3 plants, which included leaves, tree bark, and various grasses, although C4 plants were abundantly available. The diet was essentially similar to that of *Ar. ramidus*, which lived two million years earlier, and to that of modern chimpanzees, different in both cases from that of other African hominins of the same period.

Tooth analyses can reveal other aspects related to the lifestyle of ancient humans and pre-humans. Food leaves permanent isotopic traces on the teeth, allowing us to follow the migrations of our ancestors in their different environments as they are recorded in their dental and bone tissues. Metabolism induces isotope fractionation in light elements, including oxygen, carbon, and nitrogen. In the case of oxygen, the isotope ratio recorded in biominerals allows us to deduce the isotopic ratio of the ingested rainwater, assuming that the isotopic fractionation due to metabolism is known. This method was used to determine the place of origin for Ötzi, the mummy from Tyrol, on the basis of the different oxygen-18 to oxygen-17 ratios characterizing the water masses north and south of the location where the Ice Man was found.

Heavy elements such as strontium, neodymium, and lead do not exhibit measurable isotopic fractionation effects when they pass through the food chain—from soil to plants, then to herbivores, and finally to carnivores. Their isotopic concentration in teeth is a signature of local geology. For example, strontium consists of strontium-86 and strontium-87, and the concentration of the latter increases in presence of rubidium-87, which decays producing strontium-87. In carbonate-rich sediments, which have a high concentration of strontium, rubidium is essentially absent; hence there is no accumulation of radiogenic strontium-87. Carbonate sediments have a strontium-87 to strontium-86 ratio that varies between 0.706 and 0.709. Granites are much richer in rubidium; hence they are characterized by greater strontium-87 to strontium-86 ratios than carbonates.

Strontium accumulates in the tooth enamel through the food chain, thanks to its geochemical affinity with the apatite that constitutes enamel. The strontium-87 to strontium-86 ratio can thus fingerprint the geographical areas in which the individual lived during the early stages of his or her life.

For example, by comparing the strontium isotopes measured in the molar of a Neanderthal of more than 40 KYA, whose remains had been discovered in Lakonis, Greece, with those found in the soils of the surrounding regions, scientists have shown that this specimen had spent its childhood more than 20 kilometers away from the spot where it died, consistent with archaeological data suggesting that Neanderthal populations were characterized by seasonal nomadism.

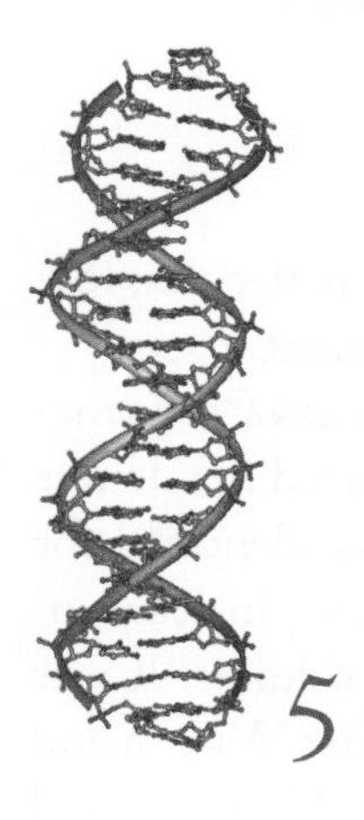

5

Reading Molecules in Fossils

Written in the Blood

In the 1960s, Luigi Luca Cavalli-Sforza, then at the University of Pavia, Italy, began a program of genetic studies to reconstruct the family tree of modern humans and the migration routes that took them to every corner of the Earth. Cavalli-Sforza's method was based on twenty-five genes associated with blood groups analyzed in fifteen indigenous populations who lived on different continents. In particular, the study was based on the analysis of proteins as genetic markers. The first evolutionary tree obtained by Cavalli-Sforza showed that the maximum genetic distance was between the groups that lived in Africa and the Australian Aborigines. He was not able to identify the origin of the evolutionary tree, although the data suggested that the first separation occurred between Asians and Afro-Europeans. The timing and routes of human dispersal remained controversial, and it was clear that the study initiated by Cavalli-Sforza was only the beginning of a long-term program. It was also clear that the solution to the human dispersal puzzle required integration of the data derived from molecules, fossils, linguistics, paleogeography, and the paleoenvironment.

In the 1980s, the advancement of technologies for DNA sequencing and the availability of powerful computers introduced the double helix into the study of population genetics. DNA was identified for the first time in 1869 by Swiss doctor Friedrich Miescher in cells isolated from pus found on the bandages of wounded soldiers. It was not until 1953, however, that American biochemist James Watson and his British colleague Francis Crick, along with Rosalind Franklin and Maurice Wilkins, reconstructed its three-dimensional double-helix structure. The existence of a genetic code responsible for transmitting biological information had been suggested in 1944 by physicist Herman Schrödinger, one of the fathers of quantum physics. The deciphering of the code could finally start in 1975, when British biochemist Frederick Sanger invented a method for sequencing DNA.

The DNA code for both the cell nucleus and mitochondria is written with four letters: A, C, G, and T, namely adenine, cytosine, guanine, and thymine. These are the nitrogenous bases aligned with the double-helix strands of the DNA molecule on a "backbone" structure of phosphate and deoxyribose. Each base on one strand is linked to a complementary base on the other strand by a weak hydrogen bond: A is always linked to T, and G to C. Mitochondrial DNA (mtDNA) mutates more rapidly than nuclear DNA, thus providing a more evident signal that can be used by geneticists to measure the genetic variations among individuals (and populations) that accumulated through time.

FIGURE 5.1
(*Opposite*) Tree of genetic diversity (mtDNA) as proposed by Cann, Stoneking, and Wilson in 1987 to show the recent African origins of our species. Horizontal lines show genetic distances. The tree shows the largest genetic divergence between sub-Saharan Africans and all the other groups, including other Africans.

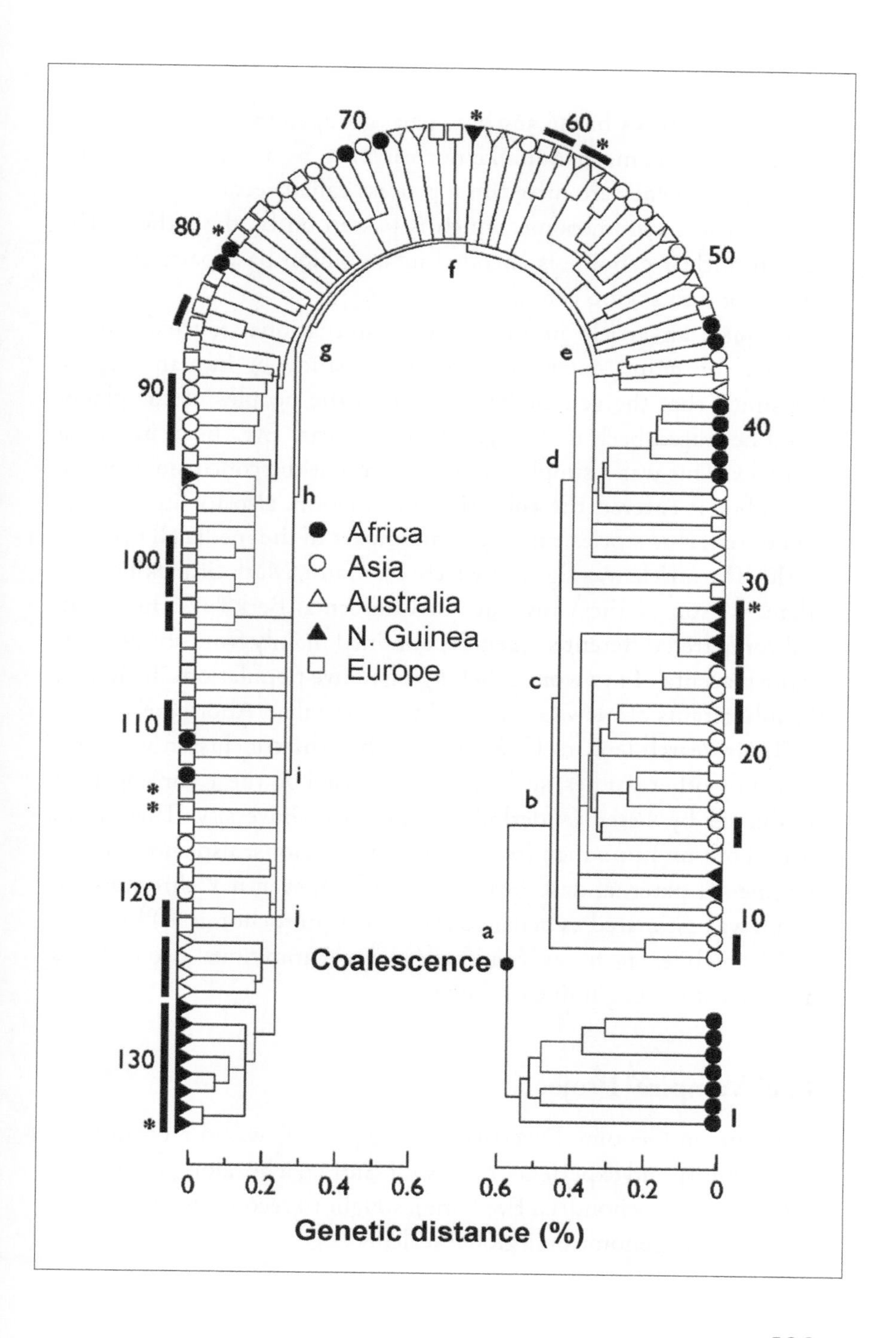

70
60
80
50
f
g
e
90
40
d
h
Africa
Asia
Australia
N. Guinea
Europe
100
30
c
110
20
i
b
120
j
10
a
Coalescence
130
1
0
0.2
0.4
0.6
0.6
0.4
0.2
0
Genetic distance (%)

Human mtDNA has 16,569 base pairs grouped in 37 genes. It is inherited from the mother, unlike nuclear DNA, in which the maternal and paternal genes are mixed in a process called recombination (except for the Y chromosome, which is passed on by the father). The human nuclear genome is encoded in 3.2 billion base pairs grouped into 20,000 to 25,000 genes.

In 1987, an article in the leading international journal *Nature* marked the grand entry of DNA in the study of human origins. It claimed that the genetic heritage of all the peoples of the planet could be traced back to a single "mitochondrial Eve," born in Africa 200 KYA. This was a simplified way to present the conclusions of the study, but it entered the collective imagination, also for its connection to religious stories, and made an impact in the media all over the world. The article was signed by Rebecca Cann, Mark Stoneking, and Allan Wilson of the University of California, Berkeley. The group had compared different sequences, obtained mainly from the analysis of the placenta of 147 women belonging to five populations, including the inhabitants of New Guinea and the Australian Aborigines.

The research lent itself to criticism, but the conclusion was confirmed by other studies, such as the one based on the Y chromosome conducted by Peter Underhill of Stanford University. This study, carried out on 1,000 men from twenty-two regions, also showed that the lines of paternal descent began in Africa, even if Y chromosome Adam was estimated as being younger than (mitochondrial) Eve.

Meanwhile, again at Stanford, a very ambitious project was launched at the beginning of the 1990s.

The "Vampire" Project

The Human Genome Diversity Project (HGDP) was the product of the ingenuity and experience of Cavalli-Sforza, who, along with Allan Wilson (of mitochondrial Eve fame), sought to reconstruct variations in the human genome on a global scale.

They organized research groups in all regions of the world, who had to collect samples from at least 100,000 individuals belonging to 400–500 populations for genetic analysis at a cost of over $30 million. A fraction of the blood samples would then be used to create cell lines to be kept indefinitely for the production of new DNA. They would also take samples of hair and saliva.

The HGDP had the scientific goal of reconstructing the origins of *H. sapiens*, comparing the data from genetics with those from linguistics and anthropology. The aim was to understand human migrations, demographic dynamics, and other factors connected with human and cultural evolution. The project had also laudable goals of human progress in general, including the fight against racism and discrimination. There were also medical objectives, and it was this aspect that associated the HGDP to the Human Genome Project, creating political problems. The Human Genome Project was launched in the early 1990s by the Department of Energy and the National Institutes of Health of the United States with the aim of completing the sequence of the human genome, which at the time was thought to be composed of at least 100,000 genes. The goal was achieved in 2001 with the publication of the results obtained by the Human Genome Organization and Craig Venter's Celera Genomics.

Australian Aborigines and other indigenous groups called the HGDP the "vampire project," because they were convinced that its objective was to use their genetic heritage for profit. They had not forgotten the attitude of the "scientists" of the nineteenth century, such as English eugenicist Francis Galton, who assigned to Aborigines, indigenous people, and blacks a very low position in the scale of human "intelligence." In any case, the HGDP created a DNA database from 1,000 individuals belonging to fifty-one populations, which was preserved at the Center for the Study of Human Polymorphism in Paris and remains accessible to nonprofit research.

This type of study was refined in the following years, providing details on demographic dynamics in different regions of the planet. In

2008, the analysis of the mitochondrial genome of 624 individuals from sub-Saharan Africa showed that, in the first 100,000 years of existence of *H. sapiens*, only a small number of maternal lines survived on the continent. This would confirm what we suggested earlier about the impact of possible environmental disasters during this period, such as the eruption of Toba in Sumatra, 74 KYA. The analyses of African mitochondrial DNA showed that in the following period forty maternal lines developed in isolation, but only two lines abandoned Africa.

Other research groups tried to follow the story of the initial dispersion of modern humans along the equatorial belt into other areas, for example in South Asia, finally reaching Sahul.

A 2009 study analyzed the mtDNA in nearly one thousand people from twenty-six indigenous tribes of India. Researchers identified seven genomes that shared two polymorphisms with the M42 haplogroup, which identified the Australian Aborigines. They also obtained a date for the separation of the Indian and Australian genetic groups: about 55 KYA, consistent with the archaeological evidence.

The Genographic Project

An ambitious global project with the aim of reconstructing the origins and migrations of modern humans was launched in 2005. The Genographic Project, funded by IBM, National Geographic, and Gateway, the computer giant, aimed to take samples for DNA analysis from 100,000 individuals of indigenous peoples from all continents (100 samples for each of the 1,000 selected populations). The study was also expected to include paleogenetic data, a topic we will discuss in more detail below.

The Genographic Project was designed to answer fundamental questions about our origins. From which area of Africa do the oldest genetic lineages of modern humans come from? What are the evolutionary mechanisms underlying the diversity of humanity today? How many waves of immigrants have populated Australia and the Americas? Did *H. sapiens* exchange genes with other species such as *H. erectus* in Southeast Asia and Neanderthals in Europe?

In planning for this ambitious project, under the direction of Texas geneticist Spencer Wells, the organizers tried to build on the lessons learned from the HGDP, including on matters of politics and public relations. Indigenous communities were kept regularly informed of all the details of the project, with rigorous protocols for obtaining approvals from ethics committees established locally in the different areas of interest. All information had to remain available to the scientific community, and there were no studies of genes of medical relevance. Finally, the preservation of cell lines was avoided because many indigenous groups oppose the idea that part of their bodies may continue to live after death.

The Genographic Project is based primarily on the analysis of mtDNA and the chromosomes X and Y. In particular, scientists cataloged genetic "markers," specific DNA sequences that define the haplogroups and keep track of the geographical dispersion of modern humans in space and time. These markers are used to follow the dispersion of the different haplogroups on timescales determined by the extent of genetic diversity among populations. The longer the separation between two human groups, the greater their genetic difference, as determined by the accumulation of mutations. Thus, genetic markers allow us to identify the common ancestors of two populations and to reconstruct the family tree.

The data of the Genographic Project confirmed on a larger scale what had been observed in previous studies: the descendants of the 200,000-year-old mitochondrial Eve separated into two groups, called L0 and L1. L0 originated in East Africa 100 KYA and later migrated through a large area of sub-Saharan Africa. Populations currently exhibiting this marker are found mainly in central and southern Africa. Later, about 80 KYA, the haplogroup L3 appeared and was the first to leave Africa, moving north. The L3 marker is found today in

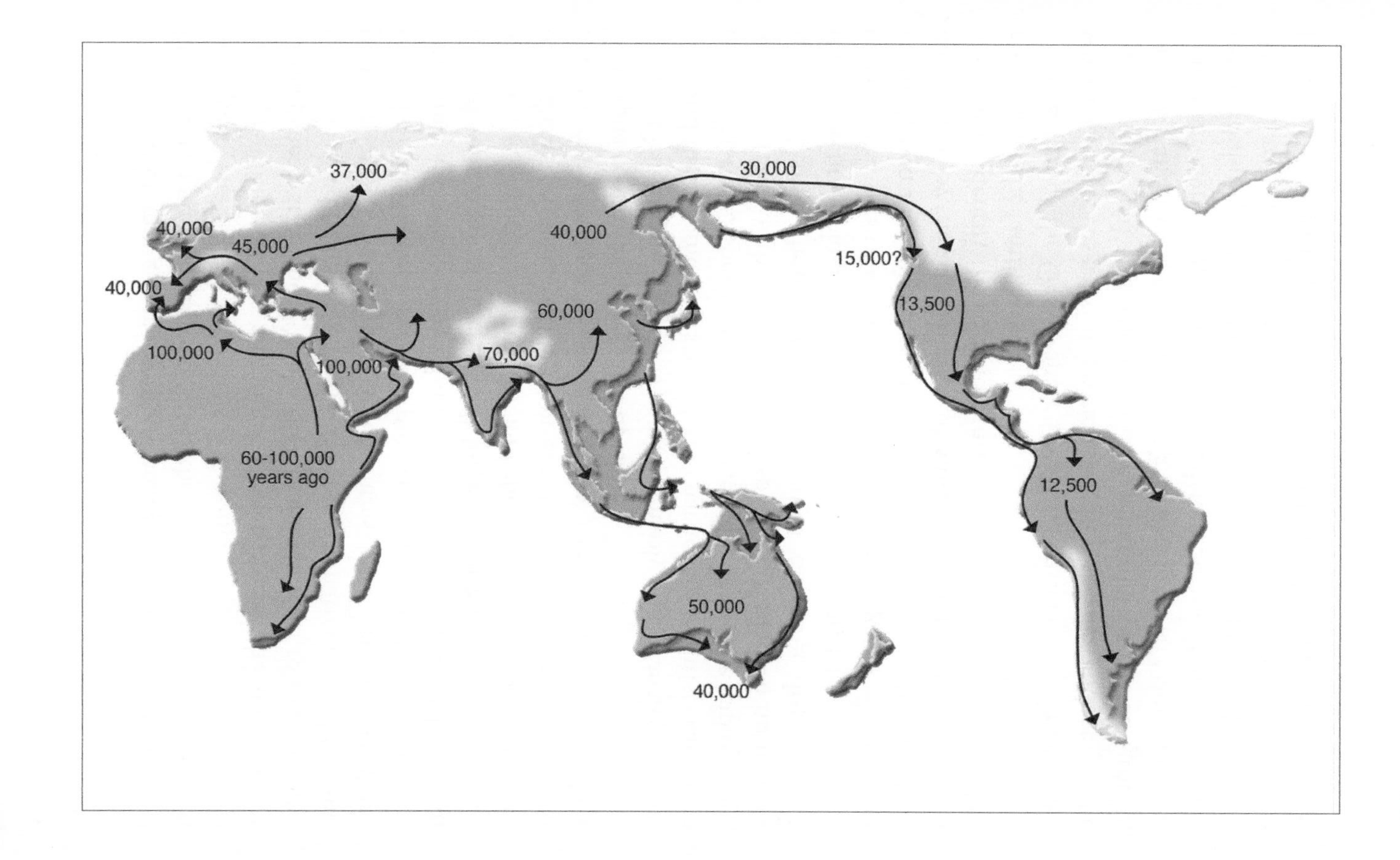
37,000
30,000
40,000
45,000
40,000
40,000
15,000?
13,500
60,000
100,000
70,000
100,000
60-100,000
years ago
12,500
50,000
40,000

the populations of North Africa and the Middle East. Haplogroup L3 split into several other groups, including haplogroup M, which appeared about 60 KYA and crossed the strait of Bab-el-Mandeb, between the Horn of Africa and the Arabian Peninsula, starting its dispersal eastward. Moving along the Indian Ocean coast, this group first reached Sunda and then—after crossing about 100 kilometers of sea—Sahul, where it eventually landed as haplogroup M42.

A similar story is told by the Y chromosome. Its oldest line of descent is defined by marker M91, which appeared 55 KYA. A branch of this group, defined by the M130 marker, was part of the group that dispersed out of Africa. In contrast, the group represented by the M242 marker dispersed into Siberia and Beringia about 20 KYA, and thence moved to the Americas.

More than half a million people have already participated in the Genographic Project, including over 50,000 individuals belonging to indigenous people worldwide. Sampling has finally begun in Australia, where Aborigines had engaged in very tough opposition for a long time. A new project is presently starting, Gene 2.0, still financed and coordinated by the groups that were involved in the Genographic Project. The new studies will identify not only the history and the area of origin of a particular genotype but also possible degrees of kinship with different human species, Neanderthals and Denisovans. Obviously, the DNA of the latter two hominins can only be found in their fossil remains.

Ancient DNA: Problems and Solutions

Until recently we could investigate human evolution in only two ways: either directly, through fossil remains and the associated paleoenvi-

---

FIGURE 5.2
(*Opposite*)Dispersal of modern humans according to genetic and archaeological evidence. Coastlines during glacial times are shown in dark gray. Redrawn from Tuniz et al. (2009).

ronments, or indirectly, by studying the current genetic variability of individuals, populations, and related species, and then attempting to reconstruct their history.

However, new approaches are now available that can extract genetic information directly from the remains of humans, animals, and plants that were alive in the deep past. There are still many technical problems to solve, but it is now possible to analyze reliably the DNA of organisms that died tens of thousands of years ago. We can finally decode the genes of our ancestors and of the infinite variety of animal and plant species that have populated the Earth and that have left suitable evidence.

When an organism dies, its molecules, including the DNA contained in its cells, begin to degrade. In particular, the nucleotides constituting the long chains of DNA break apart. Chromosomes are split into smaller and smaller fragments, until it is no longer possible to determine the order in which the fragments were connected, i.e., the DNA sequence. Under normal physiological conditions, nucleic acids decompose spontaneously as a consequence of hydrolysis and oxidation. However, whereas a living cell has enzymes that repair damaged DNA, a dead cell is no longer able to do so.

The degradation rate of DNA depends on various environmental factors, such as the average temperature of the sediments, their acidity, humidity, and other features. Only in particularly favorable conditions, such as in low temperatures or after a rapid drying of the body, is it possible to determine reliably the DNA sequence of very old fossils. Hence it is relatively easy to study the DNA of Siberian mammoths, but not, for example, that of Egyptian mummies, even if they were more recent. In essence, there are no major problems in characterizing the DNA extracted from well-preserved museum specimens that are 100 or 200 years old, but the situation is more complicated for fossil remains from the deeper past. In this case, the DNA is not only broken into fragments that may have fewer than 100 or 200 bases, but it is also present in very small amounts. This problem can be overcome by a technique, the polymerase chain reaction (PCR), which, through a series of amplification cycles, allows us to obtain a very high number of copies of a DNA fragment of interest.

However, the great sensitivity of this technique is a double-edged sword. Every ancient artifact has been manipulated and potentially contaminated from the time of its burial onward. Many individuals may have left their own DNA molecules on the sample: archaeologists, museum staff, physicists involved in the analysis of its compositions, and finally, the molecular paleoanthropologists themselves. According to an Australian ancient-DNA expert, archaeologists are the most dangerous interference, because they love to handle the bones, they get very enthusiastic about them, and, in such circumstances, perhaps they salivate a little more. It has also been shown that DNA exists in the form of aerosols even in the most controlled laboratory environment. We are all surrounded by a cloud of our own DNA! Exogenous genetic material can easily contaminate the DNA of the specimen under study and then be amplified. It is relatively simple to identify this contamination when studying animal remains (because the exogenous DNA can be identified as human), but it is much more difficult when analyzing the fossil remains of our ancestors. This is why the DNA analysis of human specimens must be based on a rigorous experimental protocol, which includes duplicate analyses by independent laboratories. It is also common practice to analyze the DNA of those individuals who might have contaminated the specimen.

We will discuss the problems posed by ancient DNA studies using as an analogy the decoding of a poem from an ancient manuscript. The full reading of the "poem" (ancient DNA) could reveal an impressive amount of detail on the art of writing, the way of life, and the emotions of people from the past. However, the DNA that we can recover from human, animal, or plant remains—an Egyptian mummy, an insect preserved in amber, or a grain of pre-Columbian maize—is, in general, fragmented into pieces not longer than 200–300 millionths of a millimeter. If these fragments were converted into pages of our ancient manuscript, they would not contain many verses. Hence, if the original poem was reporting, for example, a long and complex story, many of the more detailed and complex concepts would inevitably be lost. Only some simple information and the most basic concepts (those that can be expressed with a small number of terms) would be comprehensible.

If we want to make the comparison more appropriate to the state of conservation (or, rather, degradation) of the DNA extracted from ancient remains, we should imagine that the text in the manuscript fragment is not only reduced to a few verses, but it also faded in some points.

To further complicate the decoding of our ancient poem, broken into fragments and with some verses partly washed-out, some verses are present that do not belong to the poem itself. These passages can be in different languages: if the poem of reference was the *Divine Comedy*, we could find verses of a song by John Lennon, with words and phrases in a language that is not attributable to the Florentine language of 1300 AD, or other verses not based on the Italian alphabet (e.g., they could be in Arabic). On the other hand, there could be other more subtle "contaminations," still in Italian, but belonging to a song of Italian rock singer Zucchero Fornaciari.

In terms of our literary analogy, the ancient DNA (the *Divine Comedy*) can therefore be contaminated by modern DNA (songs in English or Arabic) that is still distinguishable. In paleogenetics, this contamination could come from bacterial or other non-human DNA. Such exogenous genetic material is easily distinguishable: even a non-expert understands that these verses do not belong to the *Divine Comedy!* The problem with analyses of ancient DNA is that subtler contaminations, such as those coming from modern humans, are difficult to identify, and this can lead to gross errors of interpretation. It is as difficult as distinguishing words in Dante's Italian from words in contemporary Italian: the danger comes from the modern rock singer!

In the following chapter we will enter into some technical details to describe the experimental methods that have been developed to recover and reconstruct the genetic material contained in ancient

Modern DNA
Nel mezzo del cammin di nostra vita
mi ritrovai per una selva oscura
ché la diritta via era smarrita.
Ahi quanto a dir qual era è cosa dura
esta selva selvaggia e aspra e forte
che nel pensier rinova la paura!
Tant'è amara che poco è più morte;
ma per trattar del ben ch'i' vi trovai,
dirò de l'altre cose ch'i' v'ho scorte.
Ancient DNA
Nel——
——selva oscura
ché la diritta via era smarrita.
Ahi quanto a dir qual era è cosa——
——è più morte;
ma per trattar del ben ch'i' vi trovai,
dirò de l'altre cose——
Contaminated ancient DNA
Nel——
Imagine all the people living life in peace...
You may say I'm a dreamer but I'm not
——selva oscura
ché la diritta via era smarrita.
Ahi quanto a dir qual era è cosa——
respiri piano per non far rumore
ti addormenti di sera ti risvegli
col sole sei chiara come un'alba
ثم الى ربكم ترجعون * ولو تر è più morte;
ma per trattar del ben ch'i' vi trovai,
dirò de l'altre cose——

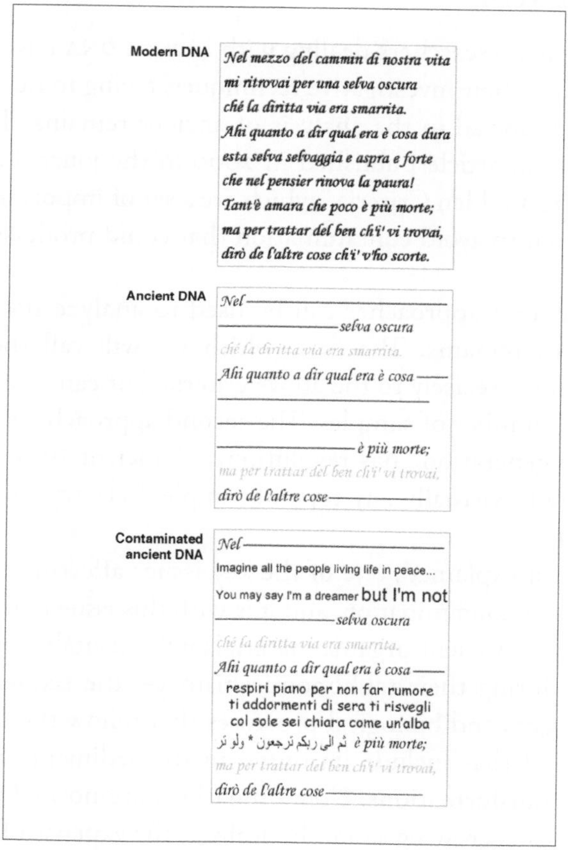

FIGURE 5.3
Ancient DNA represented by the most famous Italian poem (Dante Alighieri, *La divina commedia*). Compared to modern DNA(top), ancient DNA is missing some parts, while other parts are difficult to read because of their degradation (middle). The DNA can also show various kinds of contamination (bottom).

organic remains, with particular emphasis on the prevention and discrimination of contamination effects.

Golden Criteria

In recent years, researchers dealing with ancient DNA have attempted to standardize their investigative techniques, trying to develop a protocol to be followed in the analysis of ancient remains. These rules, described in an article published in 2000 in the journal *Science*, are known as the "Golden Criteria," which are a set of important practices to be followed to avoid contamination that could produce unreliable results.

Two different approaches can be used to analyze DNA extracted from ancient remains. The first, which we will call the "classical" method, refers precisely to the above criteria, but can be applied only to a limited number of samples. The second approach, which we will call "of new generation," has revolutionized ancient-DNA studies and is applicable to virtually any type of sample. Let's try to understand them better.

As we just explained, one of the key issues affecting the study of ancient DNA is contamination, and it is with this issue that researchers must struggle. Ancient organic materials will inevitably become contaminated during their taphonomic history—the sequence of geochemical events and biological processes that follow the death of the organism and that include its burial in the sediments and various diagenetic transformations. Contaminations are normally caused by microorganisms growing saprophytically at the expense of the cellular material of interest. During this process, the original DNA is depleted, while the contaminating DNA of the microorganism is added to the material. In the analysis of human remains, however, molecular analysis techniques (such as PCR) allow us to discriminate between the genetic component of interest and any bacterial or fungal component.

The real problem arises from contamination by current human cells (such as those produced by the epidermal desquamation of people handling the specimen). We should also add the possible contamination from equipment and reagents used for the molecular analysis by PCR. In the latter case, the same products of the reaction (fragments of current human DNA that have been amplified) become the main cause of contamination. The final result would obviously be totally unreliable.

In principle, we can always develop suitable laboratory protocols to prevent contamination from modern DNA during chemical processing. Laboratory operators should wear gloves, masks, caps, and sterile gowns. Liquid pipetting systems should be sealed from aerosols. Separated rooms should provide areas for pre-amplification and post-amplification protocols. Tools used for the extraction and amplification of target DNA should be adequately sterilized together with all the reagents, and reagents should be apportioned according to the quantities necessary to work on a single sample, thus avoiding their repeated use. Such procedures should be carried out in a special cabinet or "hood" that allows only the operator's hands inside its protected environment. Such hoods are typically flooded with a constant flux of UV rays (except of course during the execution of the experiments) that breaks up extraneous DNA. In addition, during all the steps of extraction, purification, and amplification, we should include negative controls together with the samples of interest. Such negative controls are test tubes containing all the reagents but not the original sample (e.g., bone powder) or the DNA that was extracted from it. The controls are amplified along with the samples. If any DNA is amplified in these controls, then serious doubt must be cast on the results from the actual samples. Finally, in a laboratory where modern human DNA is analyzed, the operators' DNA should be also sequenced to check whether it could have been accidentally amplified together with—or instead of—the ancient DNA.

More complicated to handle is any contamination that occurred before the specimen entered the lab. We can, in some instances, use cleanup procedures to remove and exclude, say, the surface of the sample (with abrasives if it is bone material, or by taking cores). In other cases, however, the small size of the sample, or its fragility, makes this type of operation difficult. In such circumstances, the identification of the ancient DNA can be carried out via the differential analysis of the PCR products obtained concurrently from the ancient and modern DNA, but this can be a very laborious operation.

An emblematic case of contamination due to negligent manipulation concerns Ötzi, the iceman that we have already met several times in this book. A reconstruction of the phases of recovery and preliminary inspection of the mummy, carried out at the University of Innsbruck, indicated that over thirty people had been in contact with the body in the days immediately after its discovery (September 19, 1991). Hence, it is not surprising that a 1994 study showed the presence of the DNA of many individuals in the tissues of the mummy. More recent studies on the DNA extracted from the internal organs of Ötzi, which had not previously been manipulated, finally gave reliable results.

Another controversial case concerned the DNA of Mungo Man in Australia, whom we also met earlier. The resulting DNA sequences were compared with those of thousands of extant people, including forty-five Aborigines, and with the DNA of two Neanderthals, one pygmy chimpanzee, and a common chimp. The researchers involved in the study deduced that Mungo Man's mitochondrial DNA contained a sequence that was different from all samples considered for the comparison, calling into question the validity of the out-of-Africa model. These analyses were criticized and rejected by the scientific community because the analytical procedures did not comply with the "Golden Criteria." The Aboriginal groups have recently authorized new analyses, which are still in progress.

A basic rule for the study of ancient DNA requires that the collection of the remains be done with special care, recording as much information as possible on the material of interest—including its disposition, recovery, and storage. Therefore, it is essential to collaborate with all the scientists who have worked on the same material for other kinds of studies. The primary requirement in ancient-DNA studies is thus the use of suitable experimental procedures aimed at minimizing contamination and making possible the identification of exogenous genetic material.

Reading Ancient DNA

Over the past decade there has been a strong incentive to develop new strategies and techniques for DNA sequencing, not only to complete the Human Genome Project but also to meet the needs of pharmaceutical and biomedical programs. As a result, in the last few years new approaches for high-throughput sequencing (NGS, next generation sequencing) have been developed, which can reduce both the time and costs of analysis.

The first instrument of this type, the GS20 sequencer, was developed in 2005 by 454 Life Sciences, based in Connecticut, United States, and owned by Roche. Although this tool was designed to analyze modern DNA, the company stated that the new sequencer would ensure the realization of the Neanderthal Genome Project, allowing it to "sequence a quarter of a million individual DNA fragments, taken from small bone samples, in just five hours, with a single machine."

The GS20 sequencer had been already used in 2005 to analyze the 13 million base pairs that make up the nuclear genome of a *Mammuthus primigenius* dated to about 28 KYA, paving the way for the more complicated and fascinating project on the Neanderthal genome. However, the GS20 was first used on a famous *H. sapiens*: the genome of James Watson was sequenced in just four months, at the cost of $1.5

million. Just a few years earlier, in 2001, the sequencing of Craig Venter's genome, using an earlier generation system, cost $100 million.

NGS sequencing methods are suitable for overcoming the difficulties associated with ancient-DNA analysis, providing some obvious advantages. First, it is possible to determine the state of preservation of the specimen by estimating the state of fragmentation of its DNA molecules. It is also possible to avoid multiplying any bacterial sequences, thus minimizing the generation of hundreds of thousands of undesired copies. Furthermore, the sequence lengths of the amplified copies of desired DNA are similar to the average size of the DNA fragments preserved in fossils (70–200 base pairs). Finally, the high number of molecules that are sequenced allows us to estimate the composition of endogenous and contaminant molecules. This is done by analyzing the pattern of "misincorporation" (incorrect incorporations of DNA bases during the reconstruction of the molecule due to its degradation) along with the pattern of fragmentation of the nucleotides at the ends of individual DNA molecules. In conclusion, the NGS method overcomes the problem of fragmentation, one of the main obstacles in ancient-DNA analysis, via so-called shotgun sequencing, in which the chopping of DNA into short segments of nucleotides is an integral part of the process.

Interesting data have emerged from a 2010 study in which fractions of Neanderthal DNA that were deemed to be highly contaminated were analyzed. In particular, scientists amplified the mtDNA region in which exogenous and endogenous DNA could be distinguished by comparing known sequences of Neanderthal and modern mtDNA. The analysis showed a distribution of the average length of the DNA fragments corresponding to more than 81 nucleotides for the contaminant molecules and of 45–60 nucleotides for the endogenous molecules. This can be a first clue toward distinguishing endogenous from contaminating DNA, but it is not sufficient. A clearer distinction is provided by the fact that misincorporations made up between 28 percent and 35 percent in endogenous DNA and between 4 percent and 2 percent in contaminant DNA. This pattern allows a clear distinction to be made between endogenous and contaminant DNA.

Advanced sequencing tools have had a strong impact on our ability to find answers to biological and genetic questions. Between 2005 and 2008, for example, more than one hundred papers related to the application of NGS technologies to ancient DNA studies have appeared in the scientific literature. Currently there are several technological solutions for "ultramassive" sequencing thanks to an approach that allows the parallel analysis of thousands of DNA fragments[1] reaching sequencing speeds more than one hundred times higher than previous technologies. They are being used in the 1000 Genomes Project, which has the goal of producing a map of variations in the human genome that are common to more than 1 percent of the human population. These technologies are the result of a complex combination of biochemistry, high-resolution optics, and advanced computing. Obviously, profound changes are required for the analysis, as classical statistical methods and algorithms are not sufficient to analyze the amount of genomic sequence data being generated. Advanced analytical strategies are thus required to unravel the structure, organization, and function of the genome in various human species.

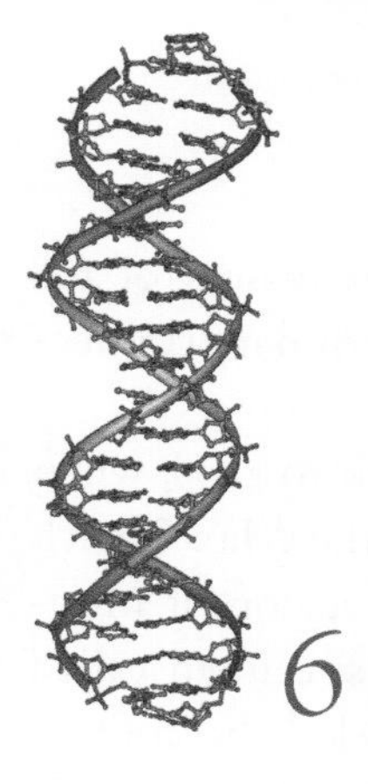

6

Stories of Molecules and Hominids

Neanderthal and Cro-Magnon

Genetic studies of anatomically modern humans who lived in Europe in the late Pleistocene and who are commonly called Cro-Magnon have focused mainly on the analysis of a non-coding region of the mitochondrial genome (hypervariable region I, HVR-I). The analysis of this molecular marker has shown that the genetic variability of the first European *H. sapiens* was not so different from that of the human populations that inhabit the same areas today. These analyses have also confirmed a strong genetic discontinuity between the Neanderthals and us.

Our knowledge of Neanderthal genetics had a significant boost from the development of ultramassive sequencing systems. Thanks to these new technologies, we can now read millions of DNA base pairs with great accuracy in just a few hours. Hence there has been an explosion of knowledge relating to both mitochondrial and nuclear DNA extracted from the fossil remains of ancient representatives of our own species and of other species of the genus *Homo*. Before describing recent research, we will sketch a brief summary of the history of ancient DNA studies and their implications for our evolutionary connections to the Neanderthals.

The first analysis of Neanderthal DNA dates back to 1997, when a research team led by geneticist Svante Pääbo of the Max Planck Institute published the first sequence of the HVR-I region of mitochondrial DNA extracted from a Neanderthal-type specimen called Feldhofer 1. The results of this pioneering study fueled the controversy on the relationship between Neanderthals and modern humans by showing that we are two genetically distinct species. It was also possible to calculate the time when the two lineages separated: the resulting date for the "most recent common ancestor" was estimated to be between 550 and 690 KYA.

Since 1997, other sequences of HVR-I were obtained from other Neanderthal remains found across Europe and northwestern Asia, including Spain (various remains from the cave of El Sidron), France (Rochers de Villeneuve and La Chappelle-aux-Saint), Belgium (Scladina and Engis 2), Germany (Feldhofer 1 and 2), Italy (Lessini), Croatia (Vindija 75, 77, and 80), Russia (Mezmaiskaya 1 and 2, the Caucasus, and Okladnikov Cave in Altai), and Uzbekistan (Teshik-Tash). The data obtained from all these remains confirmed the earliest results from the genetic analyses of Feldhofer 1, leading to the conclusion that the contribution of the Neanderthals to our genetic pool, at least in terms of mitochondrial DNA, has been either very small or nil.

The sequences of the regions HVR-I of ancient *H. sapiens* are another source of useful data on the possible contribution of Neanderthals to the genealogy of *H. sapiens*. Studies in this direction, published first in 2003 and later in 2008, concerned the remains of modern humans from Paglicci Cave in Gargano (Puglia, Italy) dated to 23–25 KYA, i.e., to the Cro-Magnon period. The results from samples of Paglicci 25 and Paglicci 12 (obtained in 2003) showed that the sequences of these individuals were indistinguishable from present-day Europeans.[1] In other words, the HVR-I sequences of the samples fit within the mitochondrial variability of populations currently living in Europe. These results thus showed that although contemporary with Neanderthals, the two humans from the Late Pleistocene had sequences indistinguishable from those of current European populations, confirming the high genetic diversity between Neanderthals and Cro-Magnons.

In spite of the strict protocols adopted to prevent contamination in all the experimental steps needed to recover the DNA from samples of Paglicci 25 and 12, the taphonomic history of these remains was not known. As we have already explained, the major problems in ancient DNA studies arise from contamination by the DNA of the operators who handled the bones during the various phases of the work (from excavation to laboratory analysis). Therefore, the results are unreliable if it is not possible to sequence the DNA of the people who came in contact with the remains.

To overcome this problem, in 2008 scientists analyzed the mtDNA of another sample from layer 23 (dated to 28 KYA) in the Paglicci Cave, which has a known taphonomic history. In this case it was possible to obtain DNA sequences from the archaeologists who discovered the bones and from the researchers who carried out the morphological and genetic studies. The DNA sequences of the sample proved to be different from those of the people involved in the study, thus demonstrating the reliability of the analysis on the samples previously taken from the same cave. The DNA sequence of Paglicci 23 is essentially similar to the sequence most frequently found in Europe, the so-called Cambridge Reference Sequence (CRS), showing also in this case that there is no difference between contemporary and ancient *H. sapiens*, and again confirming the high genetic discontinuity between our species and the Neanderthals.

Neanderthal Genetics

Only in the last few years have researchers interested in Neanderthal genetics finally been able to sequence the entire mitochondrial genome and significant parts of the nuclear DNA of this species. The latter includes almost all of the genes dedicated to defining the phenotype. In 2006, the Max Plank Institute published two major studies on the sequencing of fragments from the nuclear genome of a 38 thousand-year-old Neanderthal sample from Croatia (Vindija 80). In these studies, different research teams used two innovative methods for DNA research on the same specimen: ultramassive analysis and metagenomic analysis. This latter method is based on technologies that simultaneously analyze the genome of all microorganisms present in a particular environment—a process that helps identify new species or creates a genetic profile of known ones.

The Vindija Neanderthal (Vindija 80) was selected because it had the lowest contamination from present-day humans, even though it had been manipulated many times over the years. The problem of contamination from modern human DNA can also affect the analysis conducted on Neanderthals, since their genome is very similar to that of *H. sapiens*. The method based on ultramassive sequencing produced a large number of sequences that required a long time to be identified and interpreted. Eventually it was determined that most of the DNA sequences (79 percent) were not related to any taxonomic order and were thought to have resulted from bacteria and other microorganisms found in the soil in which the sample was discovered. Only a small percentage (6.2 percent) could be attributed to the order of Primates, also because the Neanderthal sample had DNA fragments that were quite degraded, with an average length of not more than 30 base pairs. It was, however, possible to compare these sequences with those of modern humans. Based on the results, the researchers estimated the degree of hybridization between the two species as well as the time of their divergence, and they could develop some hypotheses about the population dynamics of the Neanderthals. Assuming that the evolutionary separation between *H. sapiens* and chimpanzee occurred about 6–7 MYA, the time of divergence between Neanderthals and *H. sapiens* dates back to a period between 500 and 600 KYA, confirming the estimates obtained with the previous mtDNA analysis. Most of the differences between the two genomes are found on the X chromosome, which could mean that this chromosome was less affected by gene flow; hence, interbreeding may have occurred between *H. sapiens* males and Neanderthal females. However, these data would conflict with the absence of Neanderthal DNA in present-day humans.

The metagenomic technique was based on the hybridization of the DNA extracted from the sample with target human sequences. This allowed specific regions of the sample genome to be recovered. The extracted Neanderthal DNA was then analyzed via massive sequencing in an effort to understand the similarities and differences between the recovered Neanderthal DNA and DNA from modern humans. Applying this approach, the researchers obtained 29 of the 35 genes that were used as a target, for a total of 65,000 base pairs. From the results obtained using this method, it was estimated that the genetic divergence between our species and the Neanderthals started about 700 KYA.

The estimates of the time of divergence between *H. sapiens* and *H. neanderthalensis* produced by the two studies are basically compatible. However, some contradictions emerge around the possibility of interbreeding between the two species, since the data from the metagenomic approach do not seem to support gene flow events. The discrepancy between the two different approaches is due to the fact that the metagenomic technique allows us to analyze a smaller number of nucleotide bases but offers the advantage of focusing on more specific regions of the genome. The different conclusions drawn from the analyses of the same sample required a verification of the data, which was eventually accomplished by separating and subsequently recomparing the nucleotide sequences according to their length. It is known that the few DNA molecules that are still preserved in very old samples are extremely fragmented. Hence, sequences that are too long could be due to contamination rather than belonging to the original sample.

In fact, the re-examination of the sequences obtained using the meta-genomic method did not modify the earlier conclusion, whereas the results obtained with the ultramassive method were different. The new values for the divergence time, calculated considering only the shorter sequences, showed a strong discrepancy with the original values. It is possible that a high percentage (some 78 percent) of sequences coming from the analysis done with 454-FLX ultramassive sequencing were due to contamination. This is partly because ultramassive sequencing is unable to prevent exogenous sequences from being amplified. Based on these refined results, the observed genetic variation suggests that the ancestral population from which Neanderthals and *H. sapiens* evolved was small, maybe just some 3,000 individuals.

A 2006 study mentioned that mitochondrial sequences had been obtained from 454-FLX ultramassive sequencing, but an extensive analysis of the Neanderthal characteristics of such fragments was not carried out. A subsequent 2008 study proved the possibility to obtain the complete sequence of the mitochondrial DNA of the Neanderthal specimen Vindija 80/33.16. Subsequently, the development of new techniques of analysis such as Primer Extension Capture—a technique based on the isolation of specific DNA sequences via capture with appropriate primers followed by synthesis using DNA polymerase—has allowed the sequencing of five complete mitochondria from five different remains: El Sidrón 1253, Feldhofer 1, Feldhofer 2, Metzmaiskaya 1, and Vindija 33.25. Having established that the variability of Neanderthal mitochondrial genomes clearly falls outside the range of variation found between sequences from modern humans, it was estimated that the time of divergence between Neanderthals and *H. sapiens* was between 520 and 800 KYA, an estimate that is consistent with previous studies based only on the hypervariable region.

Who Were They?

Genetic analysis of the Neanderthals also involved the systematic search for particular genes of functional interest that have undergone adaptive changes along the human lineage, such as the melanocortin 1 gene (MC1R); FOXP2, linked to the ability to express articulated language; the ABO blood group; and TAS2R38, the gene related to the ability of perceiving bitter tastes.

MC1R is one of the genes involved in the expression of melanin; its variability involves changes in the color of skin and hair. In particular, low activity of this gene produces little pigmentation. A fragment of 128 base pairs of this gene was analyzed from two different Neanderthals from Spain (El Sidrón) and Italy (Lessini Mountains, near Verona), dated to 43 and 50 KYA, respectively. This study led to the identification of a mutation[2] totally unknown in modern humans. It was confirmed that the sequence could be considered specific for Neanderthals and that it was not a product of post-mortem DNA degradation. The phenotypic effect of this change would have resulted in low activity of the gene, leading to a light skin pigmentation and reddish hair. Such phenotypic variants are also present in human populations, but the mutations that determine them are different from those found in the Neanderthals. The two genetic variants, therefore, provide evidence of evolutionary convergence determined by similar environmental conditions.

FOXP2 is one of the genes involved in the ability to produce articulated language; its inactivation causes a reduction in facial movements and expressive abilities. Two particular variants have been established in the lineage leading to humans after the separation from the chimpanzee line, located in positions 311 and 377 in exon 7. These variants result in two amino acid changes, thus altering the structure of the protein itself. Coalescent models show that the selection of these variants was completed before 200 KYA, suggesting that the two substitutions are connected with the emergence of complex language.

It would be interesting, therefore, to understand what kind of polymorphism is associated with Neanderthals, since from their fossil remains and other evidence in the archaeological record it is difficult to determine whether they were able to communicate through complex language. A study conducted by Krause and colleagues of the Max Plank Institute led to the sequencing of the Neanderthal FOXP2 gene in two samples from the El Sidrón site in Spain, excavated under clean laboratory conditions to prevent contamination. The results showed that the regions analyzed in the FOXP2 gene were indistinguishable from those found in modern humans. In this case, since the evidence of a possible gene flow between Neanderthal and *H. sapiens* is contro-versial (as we will discuss shortly), the most parsimonious explanation is that the variant of the gene FOXP2 was fixed before the separa-tion of the two evolutionary lines. However, it is still premature to assume, based on this one genetic result, that Neanderthals (or the common ancestor between *H. sapiens* and Neanderthals) were using a complex language similar to that of modern humans, especially if we consider that this gene is not the only one implicated in this capability.

In 2008, the ABO blood group system was characterized in two Neanderthals. The high rate of polymorphism of the human ABO system seems to be correlated with the susceptibility to various diseases, since some pathogens can use antigens A or B as receptors or can benefit from the absence of both (group O). It is also believed that all the differences that distinguish the various alleles appeared after the separation of the human and chimpanzee lines. Paleogenetic analysis has allowed us to test directly the presence of ABO alleles in the two Neanderthal subjects recovered at El Sidrón—the same ones that were used for the analysis of the FOXP2 gene. Two diagnostic deletions related to group O (the O01 haplotype) were found in both samples. However, it was not possible to determine with certainty whether they were homozygous or heterozygous for this allele. The results suggest that the O01 allele was present in the common ancestor of

Neanderthals and modern humans. An alternative hypothesis could be that there was gene flow between the two species, but this seems unlikely because of the results of other genetic tests described earlier.

The expression of the TAS2R38 gene mediates the ability to perceive bitter tastes. It is known that most of the toxic substances found in plants produce bitter tastes as determined by this gene: thus the good functionality of this sense may have been important for the proper diet of our hunter-gatherer ancestors. Most of the variations in this gene relate to three polymorphisms that are located in positions 49 (proline or alanine), 262 (alanine or valine), and 296 (valine or isoleucine). These variations result in two isoforms: proline-alanine-valine (PAV) and alanine-valine-isoleucine (AVI). The first one determines the ability to perceive bitterness, whereas the second determines the absence of this capacity. In this case, the amino acid that produces the greatest effect is the one in position 49. Heterozygotes of AVI or PAV have instead a low capacity to perceive bitter tastes. In anatomically modern humans, the PAV allele has a very high frequency (75 percent) compared to 25 percent for the AVI allele. Other rare variants also exist that determine various degrees of taste perception, including A49P, which has a nucleotide substitution that causes the amino acid alanine to switch to proline. This variant is also associated with the perception of bitterness.

Knowing the different variations in humans, scientists have now analyzed a region of the gene TAS2R38 in the same two Neanderthal specimens from El Sidrón (1253). In this case, it was only possible to amplify the genetic region that determines the production of amino acid 49, probably due to the degradation of DNA across other positions. The results obtained showed that the A49P allele of the TAS2R38 gene is heterozygous, proving that this Neanderthal was able to perceive bitter tastes, even if less efficiently than homozygous individuals could have done.

The fact that *H. sapiens* and Neanderthals share the A49P allele does not imply a possible gene flow, since we are talking about the analysis of a single individual. The current presence of various sensory variants can be explained by the action of selection in maintaining a balance in the population, probably related to demographic influences and migration patterns: the data of the study show that this selection pressure was present also in Neanderthals.

Microcephalin's Gene

The microcephalin gene (MCPH1), located on chromosome 8, is one of the regulators of brain size. In humans its homozygous mutations, which lead to a loss of functionality, cause primary microcephaly, a pathology characterized by a remarkable reduction of brain volume that may occur without changes in the general neurocranial structure and without obvious external signs. The biochemical function of microcephalin is still under study, but it likely plays a key role in the proliferation of the precursors of neural cells during neurogenesis. The MCPH1 gene has been proposed as a target of positive selection (i.e., as a gene that is beneficial for selective purposes) during the evolutionary line that led from our primate ancestors to modern humans, confirming the role of molecular evolution of microcephalin in the phenotypic changes of the brain.

Positive selection is a particular mechanism of natural selection: it occurs when the latter favors a single allele, and therefore the allele frequency changes continuously in one specific direction. In a recent research project, scientists studied the G37995C locus, where a mutation from G (the ancestral allele) to C (the derived allele) (using chimpanzee DNA as an outgroup) leads to the substitution of an aspartate amino acid with histidine. The class of individuals who share this characteristic has been defined haplogroup D. The derived allele has a recent origin (between 14 and 61 KYA), but in spite of this, it shows up with high frequency, especially in the European population, though it is rare in Africa. The rapid increase in the frequency of this haplogroup over such a short period cannot be explained only as a result of demographic expansion after the dispersal out of Africa of the first modern humans: such a development also needs the intervention of other factors such as positive selection. This was verified by simulating the expansion of African human groups: without positive selection, no other demographic factors could have led to the observed differences in the distribution of haplogroup D. It is hypothesized that this variant originated about 1.1 MYA in a line of archaic *Homo* different from the one leading to modern humans, only to be transmitted later to the *H. sapiens* gene pool. The recent origin and the high frequency of the derived allele in Europe led scientists to believe that Neanderthals, who had just started sharing their territory with the first *H. sapiens*, were responsible for its transmission into the human gene pool. Such a genetic flow between the two species could have also contributed to our present genetic pool with the derived microcephalin allele.

However, the role of both Neanderthals and positive selection in the origin of the haplogroup D is still controversial. The results of simulation models—in which it was assumed that Eurasia was colonized by groups of migrants from an African population in which this mutation was already increasing in frequency—reproduced the geographic patterns of diversity observed in the distribution of haplogroup D without resorting to positive selection or to a source outside *H. sapiens*. In this regard, it is interesting to mention that another recent work that analyzed the variability of the skeletal geometry of the first anatomically modern humans suggests a remarkable geographical distribution of individuals from Africa. However, regardless of whether the genetic diversity at the microcephalin locus evolved neutrally or was determined by positive selection, the transmission of haplogroup D from Neanderthals implies that the allele must have been present in the gene pool of this species. It is for this reason that a study was recently undertaken to analyze the microcephalin gene in a Neanderthal. The findings illustrate that the ancestral allele was already present in the Neanderthal, and therefore it seems unlikely that the D allele could be the legacy of gene flow from the genetic pool of Neanderthals to that of *H. sapiens*. It is more likely that the allele originated from a mutation in the evolutionary line of *H. sapiens*. The results from the analysis of this gene thus imply that there was no contribution from Neanderthals to the genome of *H. sapiens*. Further analysis of new Neanderthal samples and the comparison of these with complete genomes of Cro-Magnon humans should reveal what might have happened between 30 and 40 KYA somewhere in Europe and whether in fact some *H. sapiens* "escapades" with *H. neanderthalensis* actually occurred and left more or less indelible traces in our genome.

Escapades with Neanderthals

We cannot yet draw firm conclusions on the contribution of Neanderthals to the gene pool of modern humans but can only provide some evidence from statistical studies based on the few data available. Mitochondrial DNA tests have shown clear differences between the two species and reveal no trace of Neanderthal characteristics in modern humans. In principle, however, genetic drift might have deleted such archaic lines from our gene pool, even if some interbreeding did occur. On the contrary, massive sequencing conducted on the same specimen detected a high percentage (30 percent) of derived alleles, which indicates anything but negligible levels of interbreeding.

Numerous studies have since been conducted in an attempt to quantify the actual contribution of Neanderthals to the genetic characteristics of *H. sapiens*. These studies show that a very small amount of interbreeding could have occurred. One of these studies assumed that Neanderthals outnumbered *H. sapiens* groups when these first dispersed into Eurasia; hence, the genetic introgression (the transfer of genes between species) would have happened very quickly. A second study is based on data from Neanderthal mitochondria (only the known sequences covering the 360 base pairs of the HVR-I region) and from numerous modern humans from various parts of Eurasia (covering a geographical range comparable to that once occupied by Neanderthals). Data on the mitochondria of ancient *H. sapiens* (Cro-Magnon) were also included. The authors simulated a wide range of demographic scenarios using a coalescing algorithm that considers mutation rates, the size and structure of populations, and population growth rates. In the end, the models that assume that present-day and ancient *H. sapiens* belong to a genealogy that is distinct from Neanderthals better reflect our current genetic variability compared with models that assign *H. sapiens* to the same mitochondrial genealogy as the Neanderthals. The authors thus concluded that the maximum level of gene flow between Neanderthals and Cro-Magnon humans was 0.001 percent per generation, an order of magnitude lower than estimated in previous studies that did not include data on Cro-Magnoids.

The Neanderthal's Genome

In the May 7, 2010 issue of *Science*, a team led by Svante Pääbo published a draft sequence of over four billion nucleotides of the Neanderthal genome: about 60 percent of the entire genome. (It is assumed that Neanderthals had a DNA similar to our own, with 6.4 billion diploid nucleotides). The DNA was extracted from the fossil remains of three Neanderthals discovered in the Vijndia Cave in Croatia, which lived between about 38 and 42 KYA.

The genome sequence confirmed that the Vijndia individual was a Neanderthal male. The sequence was then compared with that of five individuals of our own species from different continents. The results of the research, surprisingly, showed that some regions of the Neanderthal's genome are homologous with our genome. Among these, there are some peculiarities probably typical of the genus *Homo*, such as features that affect skin color (already observed in a previous study), sweating, the roots of hair follicles, and the tongue's taste buds—the latter attributable to the family of repetin proteins expressed by the gene RPTN. This research also highlighted mutations important for cognitive development and that contribute in humans to diseases such as Down syndrome (DYRK1A), schizophrenia (NRG3), and autism (CADPS2, AUTS2). Another aspect relates to the THADA gene found on chromosome 2, which is associated with type 2 diabetes and leads us to hypothesize that Neanderthals and *H. sapiens* had different kinds of metabolisms.

Another peculiar characteristic is the differences between the RUNX2 genes in the two species. In humans, this gene is associated with a wide spectrum of abnormalities related to development, including the deformation of the clavicle and of the rib cage. This is definitely an interesting aspect, given that Neanderthals have an inverted bell-shaped rib cage and a collarbone that is very different from ours. Finally, the SPAG17 gene, which encodes an important protein that regulates sperm motility via the movement of the flagellum, is different between the two species.

Another perhaps more controversial result that emerged from the study concerns one of the issues that anthropologists have been debating for a long time: did modern humans interbreed genetically with Neanderthals? As we have seen, that possibility had been excluded following initial molecular anthropology studies. However, more recent results obtained from genetic tests and biostatistics showed that it was not possible to rule out a modest degree of genetic mixing, estimated at about 120 interbreeding events throughout the entire coexistence between the two species. Recall, though, that recent data based on the analysis of mitochondrial genomes and sophisticated simulations still exclude any type of mixing. In any case, data from the nuclear genome do show that the genome of human populations in Europe and Asia, in fact, may contain between 1 percent and 4 percent of genes from Neanderthals, whereas African people do not possess such variants. This contradictory evidence could be explained with the hypothesis that interbreeding occurred mainly between Neanderthal males and *sapiens* females (considering that mtDNA is inherited via the maternal line)—even though the contrary could have happened, according to a recent study on the Italian Neanderthals from Monti Lessini.

This result suggests that the first *H. sapiens* to migrate out of Africa arrived in the Middle East and interbred with Neanderthals. Later, those groups of modern humans multiplied enormously, and this would explain why, starting from a few interbreeding events, we came up to a maximum of 4 percent of Neanderthal genes in our genome, whereas there is no trace of our genes in the Neanderthal genome (also because the Neanderthal population was undergoing a dramatic contraction after the dispersal of modern humans into Eurasia).

To this "demographic" explanation we can add another, as Svante Pääbo and other colleagues involved in the *Science* study recognize. One might assume that the populations of the species (i.e., *H. heidelbergensis*) that was ancestral to both Neanderthals and modern humans had a number of sub-structures with different combinations of genetic and morphological traits that were inherited by their descendants, remaining fixed in the different lineages: this could erroneously suggest an interbreeding among the descendants that did not actually occur.

Strange Fossils from Denisova

In the summer of 2008, during an excavation in the cave of Denisova in Siberia, Russian researchers found amid a collection of stone tools attributed to Neanderthals or early *H. sapiens* the fragment of a bone, probably a phalanx.

At first the bone fragment was considered a finding without interest, possibly left by a Neanderthal or an ancient representative of our own species who lived in the cave between 30 and 48 KYA, and therefore it was set aside for later investigation. The analysis eventually carried out by Johannes Krause of the Max Planck Institute (in Leipzig) created havoc among paleoanthropologists. The DNA, which had been sequenced with the new methods of ultramassive analysis, did not match either the genetic material of the Neanderthal or that of modern humans who lived in the same area at the time.

In fact, the molecular data revealed a very surprising discovery: the phalanx could have belonged to an unknown extinct human who had migrated out of Africa after the first dispersal of *H. ergaster*, about 1.8 MYA. According to the genetic sequence, the individual from Denisova arrived in Eurasia much earlier than the Neanderthals ancestor—*H. heidelbergensis*, present in Eurasia from about 600 to 200 KYA—and *H. sapiens*, who spread into Eurasia during the Upper Pleistocene. In fact, the comparison between the complete sequence of the mitochondrial genome of the Denisova fossil and those of 54 present-day humans, a modern human of 30 KYA found in Russia, and six Neanderthals, has clearly indicated that the new sequence definitely lies outside the genetic variability of our species and of the Neanderthals. The Neanderthal mitochondrial genome differs from ours, on average, by 202 nucleotide positions, whereas that of the Denisova finding differs by 385 base pairs. Such a difference would imply that the evolutionary line of the Siberian ancestor separated about 1 MYA, long before the divergence between modern humans and Neanderthals. Caution is obviously a must, but if future studies confirm these conclusions, it would be the first identification of an extinct human species through DNA analysis alone. The Denisova individual seems to be a new type of hominin, but does he belong to a new species of the genus *Homo*? The geological layers in which these fossils were discovered have revealed both Middle Paleolithic and Upper Paleolithic lithic instruments, so it is not clear what culture they belong to and who might have produced these tools.

When the genome of the Denisova specimen was compared with that of Neanderthals and modern humans from different parts of the world, it was revealed that the DNA of the new species shares a large number of gene variants with modern populations indigenous to Papua New Guinea, indicating that some interbreeding took place between the ancient population of Denisova and the ancestors of Melanesians. A comparative analysis of the genomes that included samples from the populations of New Guinea and Bougainville Island revealed that at least 4–6 percent of the Melanesian genome originates from the ancient Siberian population. This observation leads to the hypothesis that the Denisovans may have dispersed into Asia during the late Pleistocene. These genetic analyses thus suggest that the evolution of modern humans and their closest relatives entailed a more complex scenario than previously thought. Some researchers believe it is likely that an ancestral group left Africa between 300 and 400 KYA, evolving rapidly into two different taxa: the Neanderthals, who dispersed into Europe, and a population that moved further east, who became the Denisovans. Therefore, when modern humans left the African continent, starting at least 120 KYA, they met both

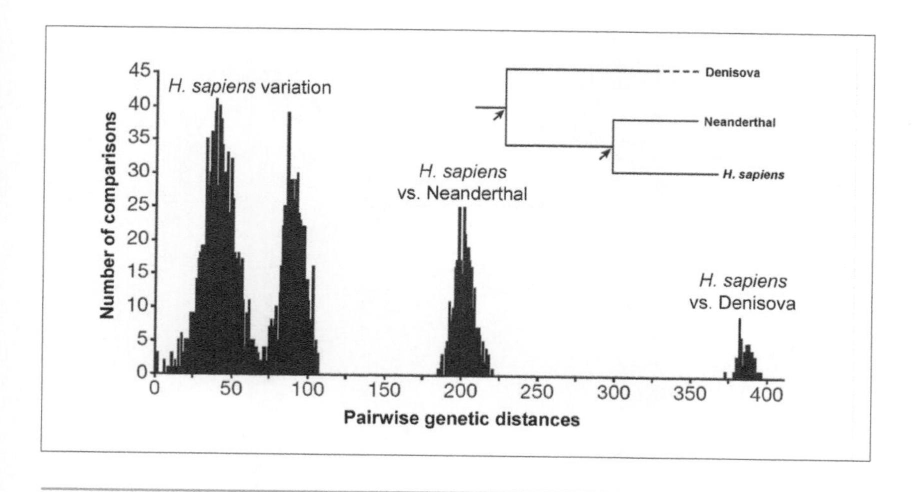

FIGURE 6.1
Comparison between complete mitocondrial genomes of *H. sapiens*, *H. neanderthalensis*, and Denisovans. The Denisova specimen was genetically different from both the Neanderthals and modern humans. The genetic difference between modern humans and Denisovans is almost double the observed difference between *H. sapiens* and *H. neanderthalensis*.

the Neanderthals and the Denisovans. Genetic traces of interactions with Neanderthals are found in all modern non-African populations, whereas the contact with Denisovans is found exclusively in Melanesian populations. At the moment it is not clear when, where, and to what extent the interbreeding would have taken place.

Although our genetic exchanges with other human species threaten the concept of biological species, various studies have shown that hybridization between related species is not uncommon in nature: evidence for hybridization is indeed being found in a growing number of primates and other vertebrates. We are still somehow far from understanding what makes us exactly who we are today.

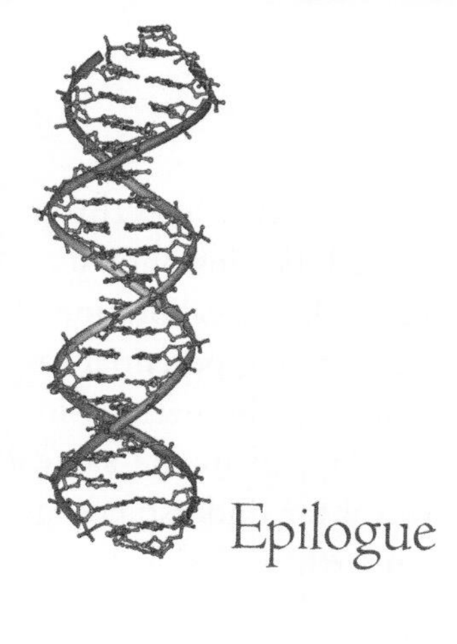

Epilogue

The peculiar genome of the Denisovans was deemed sufficiently similar to the previously sequenced genomes of *H. sapiens* and *H. neanderthalensis* to be attributed to an umpteenth new member of the genus *Homo*. Among the more recently discovered varieties of our genus from which we expect to be able to extract DNA, the genome of the diminutive human from the Indonesian island of Flores (*H. floresiensis*) is still missing, but scientists have not given up. The aim of these studies is to reveal the derived characteristics of our genome that are unique to our own species, even if it is still difficult to predict the consequences of genetic changes on functional (and pathological) aspects of our body. In addition, as we have tried to explain in this book, the biological secrets of extinct human species can be revealed by illuminating their remains with synchrotron light, X-rays, and other particles, which can provide information on the microstructure of their teeth and other skeletal remains such as the hyoid bone, kneecaps, pelvis, and ear bones, or can reveal the internal surface of the cranium.

We have also examined the increasing accuracy with which chronologies marking the human history are measured. For instance, using new methods of radiocarbon analysis and the latest calibrations of radiocarbon ages (which now push back their dating potential to more than 50 KYA), geochronologists are reviewing the periods in which modern humans and Neanderthals overlapped in the same territories of Eurasia. This information is critical for understanding the role of our species in the demise of Neanderthals. Using luminescence dating and other methods, it is also possible to synchronize the precise timescales for archaeological findings that define the trajectories of human dispersals during the last 100 thousand years with the calendars based on mutations that have occurred in both nuclear and mitochondrial DNA; these mutations can also be related to past climatic and environmental changes, which are the main fuel for the engine of natural selection and, ultimately, for its action on our biological and cultural evolution (although the effects of genetic drift and sexual selection cannot be neglected).

In any case, there is still a long way to go before we can understand the reasons behind our species' dramatic expansion—both geographically and population-wise—in parallel with the development of our cultural complexity and with the gradual decline, and ultimate extinction, of the little people of Flores, the Denisovans, and the Neanderthals, who had lived for millennia side by side with our direct ancestor, *H. sapiens*.

The road to uncover all the mysteries of human evolution is still long, but the journey has begun.

Notes

Chapter 1

1. In the following we use the designation *H. sapiens* and similar abbrevia-tions for all species of the genus *Homo*, as for example *H. erectus* or *H. neander- thalensis*; similar forms are used for the most common genera of hominins.

2. We use here the notation MYA for millions of years ago and KYA for thousands of years ago.

3. In *The Descent of Man and Selection in Relation to Sex* (1871).

4. Moreover, in many of our most ancient ancestors and extinct relatives, bipedalism was linked to other forms of locomotion in the forest (so-called arboreal locomotion); we must therefore distinguish forms of "optional bi-pedalism" (of the australopithecines and more archaic forms) from the "forced bipedalism" of the genus *Homo*, including *H. sapiens*.

5. We refer to the portion of mandible found in Chad and called *A. bahrelghazali* and to the skull found in Kenya together with other findings and attributed to the genus *Kenyanthropus* (*K. platyops*); both of these hypothetical additional taxa have been dated to about 3.5 MYA.

6. The technical term used for this type of phenomena is "exaptation," previously referred to as "pre-adaptation."

7. In this case we talk about "paleontological species."

8. Such as *H. erectus*, the human species distributed in the Far East that could have survived, at least in Java, until quite recent times (around 100–200 KYA).

9. Tentatively attributed to another species: *H. georgicus*.

10. The producers of the film on the character created by J.R.R. Tolkien have recently prevented the use of the name Hobbit as a nickname for *H. floresiensis*.

11. Not even some varieties of the genus *Homo* that also lived during the Middle Pleistocene but had much more ancient origins: in particular, *H. erectus*.

Chapter 2

1. This was done after having concentrated the carbon using various chemical treatments such as combustion (for bones and wood) and hydrolysis (for shells and other carbonates).

2. Using gas counters, in which beta particles release their energy-producing ionization, or liquid scintillators, in which they produce light pulses.

3. The periodic oscillations that occurred in the past, with periods of about eleven years, correspond to the variability of the solar magnetic field. The oscillation with a period of about 8 KYA corresponds to changes in the Earth's magnetic field.

4. Also in these cases, the calibration curve is built by dating these systems independently (e.g., through thorium-230/uranium-234 dating).

5. For many archaeological studies in Africa and other areas with a warm climate, human and environmental remains are not preserved in ice, and only isotopic analyses can provide the required information.

Chapter 3

1. Like our prolonged periods of growth and development (almost twice those of chimpanzees) and the premature birth of our young.

2. Actually, Horace-Bénédict de Saussure (1740–1799), a Swiss aristocrat and naturalist, was the first to recognize that erratic boulders had been moved great distances by ice.

3. Precipitation, and therefore ice, contains water with a low concentration of oxygen-18. During glacial periods, when a large amount of oxygen-16 is stored in glaciers, the concentration of oxygen-18 in ocean water is high.

4. This list refers to some extinct species of Australian megafauna: *Diprotodon optatum, Thylacoleo carnifex, Megalania prisca, and Genyornis newtoni.*

5. This is explained by the fact that cosmic rays, which produce new radionuclides via nuclear reactions with the atmosphere, are shielded by the Earth's magnetic field; when its intensity decreases, the flow of particles increases, and hence the production of beryllium-10 also increases.

6. The aragonite and calcite minerals are calcium carbonates with different crystalline structures.

7. Nine species (in six genera) of flightless birds endemic to New Zealand, the largest of which was 3.6 meters tall.

Chapter 4

1. Furthermore, the availability of X-rays with a specific energy (obtained with special devices called monochromators) minimizes "beam hardening," an effect that occurs with conventional X-ray sources (which produce X-rays with a spectrum of different energies). In this case, the X-rays with lower energy are absorbed by the thickest part of the sample under analysis, whereas a thin superficial part is traversed by a higher flux of X-rays. Hence, in tooth imaging, the enamel will appear brighter, leading to wrong conclusions about the thickness of the enamel itself. In addition, detectors are more efficient in the detection of "soft" (low-energy) X-rays, and this effect amplifies beam hardening. The use of high-energy monochromatic X-rays, available with synchrotrons, eliminates these effects and improves image resolution.

2. The X-ray source has a focal spot size of the order of one micron or less.

3. It has also been shown that by increasing the distance between the test object and the detector, phase contrast methods can be used with these systems.

4. One micron equals one thousandth of a millimeter.

5. On the other hand, australopithecines have semicircular canals similar to apes.

6. The carbon-13 to carbon-12 isotope ratios are measured in the tooth enamel with a mass spectrometer. The CO_2 produced from an enamel sample of two milligrams taken with a dentist's drill is made to flow in a chamber where it is ionized by electrons emitted from a hot filament. Accelerated through a magnetic field, the positive ions follow different trajectories according to their mass. The carbon-12 and carbon-13 beams reach the respective charge collectors, which measure the concentration of each isotope. Since mass spectrometers do not provide absolute estimates, we must evaluate the isotopic ratios by comparing our measurements with a calcium carbonate sample used as an international standard.

Chapter 5

1. The Genome Sequencer FLX Titanium-Roche/454 Life Sciences, the Illumina/Solexa Genome Analyzer, and Applied Biosystems SOLiD TMSystem are the most commonly used.

Chapter 6

1. Paglicci 25 and 12 are two Cro-Magnoid remains of a woman and a child, respectively, dated to about 25 KYA and found in the Paglicci Cave in Puglia.

2. This mutation is characterized by a transition that causes an amino acid to change from arginine to glycine.

Glossary

Accelerator mass spectrometry (AMS) Analytical technique used for measuring low-level cosmogenic radionuclides such as carbon-14, beryllium-10, and aluminium-26. AMS is based on the use of an ion accelerator as one component of a high-sensitivity mass spectrometer. Ion detectors allow the identification of the nuclear mass and atomic number of the atoms of interest after they have been extracted from the sample, ionized, accelerated to high energies, and stripped of some of their external electrons, and after their mass and charge have been filtered through magnetic and electrostatic deflections. We can use AMS to measure isotopic abundances as low as 10^{-16}, which in the case of carbon-14 correspond to ages of more than 60 thousand years.

Acheulean Lithic technology of the Lower Paleolithic, characterized by bifaces and cleavers. Widespread in Africa and Eurasia from about 2 MYA to 150 KYA, it is associated with *H. erectus, H. heidelbergensis*, and early *H. sapiens*.

Adaptive radiation Diversification of biological forms starting from a stem common ancestor.

Albedo The fraction of solar energy reflected from the Earth's surface back into space.

Allele One of two or more variants of the same gene at a specific chromosome location.

Amino acid One of the basic constituents of macromolecules (proteins) that make up living organisms. After an organism's death, most of the organic matter decomposes, but proteins such as the keratin of hair, the collagen of bones, the dentine of teeth, and other biopolymers can be preserved for thousands of years.

Amino acid racemization (AAR) Dating method based on the racemization of specific amino acids. In proteins of living organisms, amino acids are linked into peptides as left-handed molecules (amino acids L), a phenomenon that has been correlated with enzymatic reactions. Following the death of the organism, enzymatic reactions that kept the imbalance in favor of the left-handed L form cease, and the amino acids are converted progressively from left-handed to right-handed (amino acids D). This chemical process is called "racemization" and, due to its slow progression, can be used as an archaeological clock.

Anthozoans Class of marine organisms that gather in colonies and produce the calcium carbonate present in the coral's skeleton.

Apatite Biomineral cristalline material composed of calcium phosphate found in the tooth enamel.

Ardipithecus ramidus The etymology of this word can be found in the Afar language (spoken by the people in the Ethiopian region). *Ardi* means "ground" and *ramid* "root"; therefore this would be the "ape that lived on the ground, at the roots of the human lineage."

Argon-40/argon-39 Dating method based on the radioactive decay of potassium-40 into argon-40 ($T_{1/2}$ = 1,248 million years) in volcanic materials. The nuclide daughter, argon-40, is not present in materials that have just erupted from volcanoes, but it accumulates thereafter as the parent nuclide, potassium-40, decays. After heating the specimen to extract the trapped gas, the concentration of argon-40 is measured using a mass spectrometer. The concentration of potassium-40 is obtained by irradiating the sample in a nuclear reactor, where neutrons interact with potassium-39 producing argon-39. Finally the mass spectrometer simultaneously measures argon-39 and argon-40, providing a date for the material. Ages in excess of 4 billion years can be measured with this method.

Atom The smallest component of a chemical element. It consists of a central nucleus, composed of two types of particles of similar mass: neutrons (with neutral electric charge) and protons (with positive electric charge); smaller particles, the electrons, move around the nucleus following the laws of quantum mechanics. Almost 2,000 times lighter than the protons, electrons carry a negative charge. The number of protons—which, in a neutral atom, is the same as that of electrons—characterizes the chemical element (6 for carbon, 7 for nitrogen, etc.). Each element can have different nuclei, which take the name of isotopes, characterized by a different number of neutrons. The unstable (radioactive) isotopes are called radioisotopes.

Aurignacian One of the earliest cultures of the Upper Paleolithic, associated with modern humans in Europe and named after the first site where its remains were discovered, at Aurignac in the Pyrenees. Archaeological sites related to this culture date back between 45 and 35 KYA.

Australopithecus Genus of the tribe Hominini, which includes (for many taxonomists) four species: *A. afarensis, A. anamensis, A. Africanus,* and *A. Sediba,* plus other species that are more questionable. They lived between about 4.1 and 1.5 MYA.

Bases The four molecules—adenine (A), guanine (G), cytosine (C), and thymine (T)—that encode the DNA information. The two DNA strands are held together by the hydrogen bonds of C with G and A with T.

Beringia An ice-free passage that connected northeastern Siberia to western Canada across the Isthmus of Bering. It emerged from the sea about 13 KYA, where today the Bering Straight is located.

Beryllium-10/aluminium-26 A dating method, also called cosmogenic dating, based on the decay of beryllium-10 and aluminium-26. Sediments and rocks on the Earth's surface are exposed to the bombardment of secondary cosmic rays, and new nuclides are produced in the quartz crystals of rocks and surface sediments, including beryllium-10 ($T_{½}$ = 1.39 million years) and aluminum-26 ($T_{½}$ = 717 thousand years). When the sediments are subsequently buried at a depth of a few meters, and hence shielded from cosmic radiation, the production of new radionuclides ceases. Beryllium-10 and aluminium-26 decay and their isotopic concentration decreases with time, working like a clock that measures the time elapsed since the last exposure to cosmic rays. The isotopic concentration of beryllium-10 and aluminum-26 can be measured efficiently with AMS.

Beryllium-10/beryllium-9 Dating method based on the decay of beryllium-10, produced by high-energy cosmic rays that bombard the atmosphere. The radionuclide is subsequently incorporated into aerosol particles and transferred to the Earth's surface by precipitation. Finally, beryllium-10 is incorporated into the sediments, where it decays. Hence, the beryllium-10/beryllium-9 ratio—assuming that the stable beryllium-9 that is present in traces in the sediment was homogenized with beryllium-10—functions as a geological clock.

Beta particles Electrons moving at a speed comparable to the speed of light, emitted during the radioactive decay of certain radionuclides (e.g., carbon-14).

Biface Flat stone tool with an edge made by working both sides of a core.

Breccia Sediments that have been calcified by filtering lime solution.

Broca's area Part of the hominin brain, located in the left hemisphere, related to language. Individuals with damage in this area have speech difficulties.

Burin Stone tool that looks like a chisel, probably used for carving wood and bone.

C3 Photosynthetic mechanism used by trees, shrubs, or grasses such as wheat; the first compound of the photosynthesis is a molecule with 3 carbon atoms.

C4 Photosynthetic mechanism used by plants adapted to drought and able to retain water (i.e., corn and sugar cane); the first compound of the photosynthesis is a molecule with 4 carbon atoms.

Cambridge Reference Sequence (CRS) Mitochondrial DNA sequence to which all scientists who study human mitochondria refer for comparisons.

Cell line Cultured cell obtained from primary sampled tissues.

Cenozoic The geologic era from 65 MYA to the present.

Chromosome DNA structure that contains part of the nuclear genome. Diploid human cells have 46 chromosomes.

Cladistics Classification system based on phylogenetics, or the evolutionary relation among groups of organisms based on common ancestry.

Clovis culture Culture named after the town in New Mexico where archaeologists discovered, in 1934, typical Paleolithic artifacts to make throwing weapons. Similar objects, consisting in fluted projectile points created by using bifacial percussion flaking, were also found in other North American sites.

Coalescent theory In genetics, a theory that states that all the genes or alleles of a given population are ultimately inherited from a single ancestor, known as the *most recent common ancestor*.

Computed microtomography (microCT) X-ray imaging method based on microfocus X-ray sources or synchrotron radiation. High-resolution 3D images—from sub-micron for small mm-size objects to several microns for 10–20 cm objects—are generated from a large series (typically 1,000–4,000) of radiographs using specific mathematical algorithms for slice reconstruction. Micro-CT images are produced rotating the specimen by 180° around the axis that is perpendicular to the incident rays. The transmitted beam is recorded during rotation with a detection system generally consisting of a scintillator (for converting X-rays into photons of visible light) and a CCD camera. Other detection systems are also used.

Convergence The parallel development of the same trait by distinct evolutionary lineages, either randomly or following similar selection pressures.

Core A sample collected with a cylindrical tube inserted vertically in the ground, sea sediments, or ice and extracted to preserve the temporal sequence of the material deposited over time. In archaeology, a core is a piece of stone from which smaller flakes are removed: both the core and the flakes can be used to produce tools.

Cosmic rays Cosmic rays are composed of high-energy particles (mostly protons, i.e., hydrogen nuclei) generated from exploding stars in the galaxy (supernovas). They were discovered in 1912 by the Austrian scientist Hermann Hess (who received the Nobel Prize in 1936). Nuclear reactions of cosmic protons with the atmosphere cause the emission of muons, neutrons, and other particles, which then bombard the lithosphere.

Cosmogenic radionuclides Environmental radioactive nuclides, such as carbon-14 and beryllium-10, produced by nuclear reactions of cosmic rays with the atmosphere and lithosphere.

Cranium The braincase and bones of the face. Cranium and mandible make up the skull.

Cretaceous The last geologic period of the Mesozoic era, from 145 to 65 MYA.

Cro-Magnon European modern humans of the Upper Paleolithic. The name derives from the skeletal remains of five individuals found in 1868 in the Cro-Magnon rock shelter at Les Eyzies-de-Tayac (Dordogne, France). Together with several other fossils discovered in Europe and belonging to our own species (*H. sapiens*), these specimens were known in the past as "Cro-Magnonoids."

Cuddie Springs Archaeological site in New South Wales, Australia, where stone tools and fossils of extinct megafauna have been found.

Culture Information stored in the human brain that affects the behavior of an individual; it is transmitted by social interactions, teaching, imitation, etc. It consists of a "cognitive package" that includes the use of symbolic and self-conscious thought and of complex language, and the capacity of producing sophisticated tools.

Deciduous dentition The first set of teeth that appear before the permanent dentition.

Defect Absence of electrons in the crystal lattice, which acts as a trap for the electrons freed by natural ionizing radiation.

Deoxyribonucleic Acid (DNA) The molecule that carries hereditary information in all living organisms. It is made of a double helix of two complementary strands of nucleotides. See also **Base**

Derived character A new trait developed in a more recent organism that is retained by descendants but is absent in the older ancestry.

Diagenesis The process by which a sediment is transformed into a sedimentary rock by physical, chemical, and biological alterations.

Diploid Refers to cells containing two copies of the genome.

Electron spin resonance (ESR) Dating method based on the accumulation of effects due to the interaction of natural ionizing radiation with crystals, such as tooth apatite. Pairs of electrons occupy the same orbital, but they may become unpaired when an individual electron remains trapped in the defect of the crystal lattice as a result of the energy released by the absorbed radiation. In the ESR method, the spin of these individual electrons is aligned by a strong external magnetic field and the sample is irradiated with microwaves. The amount of microwave energy absorbed allows determination of the total number of unpaired electrons and, consequently, the age of the sample, up to more than 500 KYA. However, it is essential to have a good estimate of the annual dose of radiation absorbed by the sample material during its burial in the ground.

Encephalization The increase of brain size relative to body size.

Endocast The cast of the inside of the braincase, corresponding to the external brain structure.

Eocene The second epoch of the Paleogene period, between 56 and 33.9 MYA.

Evo-devo It stands for "evolutionary developmental biology." It is a relatively new discipline that results from the merging of molecular, developmental, and evolutionary biology, with the aim of studying the evolution of developmental processes of different organisms.

Exaptation Change in the function of a trait during evolution.

Exon A sequence of DNA that codes information for protein synthesis that is transcribed into messenger RNA.

Family A taxonomic division higher than genus and lower than order. Humans belong to the family of Hominidae, to the genus *Homo*, and to the order of Primates.

Feldhofer 1 Neanderthal type specimen discovered in 1856 in the valley of Neander, not far from Dusseldorf, Germany.

Fission track Dating method based on the accumulation, in volcanic minerals, of the effects produced by the spontaneous fission of natural uranium-238 (present in traces in these materials). The fission fragments produce microscopic traces in crystals such as apatite, which can be preserved for millions of years if the rock stays below 120°C. Tracks are made visible with a hydrofluoric-acid attack, and counted under a microscope. The density of the tracks provides the time elapsed since the last volcanic eruption, but only if the concentration of uranium is known. This is calculated by irradiating the sample in a nuclear reactor, where neutrons induce the fission of uranium-235. Ages of more than 1,000 million years can be evaluated using the fission track method.

Gene Part of the chromosome that produces a recognizable effect on the phenotype.

Gene flow Transfer of genes from one population to another as a result of interbreeding.

Genetic distance A measure of the evolutionary relationship between two populations. The concept was introduced by Cavalli-Sforza and Edwards in the 1960s using the differences between the percentage frequencies of a gene form. The distance was then averaged over a large number of genes by using appropriate mathematical algorithms.

Genetic drift Change in gene frequencies due to random factors, not adaptation, mostly in relation to sampling variation after drastic reductions in the number of individuals in the population.

Genetic markers Genes whose position in a chromosome can be used to identify other genes in a nearby position.

Genetic pool Set of all alleles of the entire set of genes that belong to all the individuals that make up a population at a given time.

Genotype The combination of alleles that characterize an individual.

Genus Taxonomic division lower than family and higher than species.

Geologic clocks Radionuclides most commonly used in archaeological dating, such as radiocarbon-14, beryllium-10, aluminum-26, potassium-40, and uranium-238; they are analyzed using particle accelerators or other advanced techniques, including laser-based microprobes.

Half-life Time after which half of the radionuclides present at the initial time have decayed (represented with the symbol $T_{1/2}$).

Haplogroup A set of mtDNA or Y-chromosomal haplotypes defined by specific genetic markers.

Haplotype The combination of alleles defined by a set of polymorphic markers that are inherited together.

Heinrich events Climatic episodes during the last Ice Age, characterized by sudden drops in global temperatures of 2°C–3°C, which lasted only a few decades. One hypothesis is that they occurred when huge icebergs broke off the North America Laurentide ice sheet, which had become unstable after growing to con-tinental proportions. They occurred at the following times: H1 at about 16 KYA, H2 at 22 KYA, H3 at 30 KYA, H4 at 38 KYA, H5 at 45 KYA, and H6 at 65 KYA.

Heterozygote A locus in which the two alleles of any diploid cell are different.

Holocene The most recent geological epoch, which began 11.7 KYA.

Hominids Members of the family Hominidae that includes four extant genera: *Pan* (chimpanzee), *Gorilla* (gorillas), *Pongo* (orangutans), *Homo* (humans). Also includes previous genera of extinct bipedal apes belonging to the human lineage, such as *Australopithecus* and *Ardipithecus.*

Hominins Members of the tribe Hominini or the sub-tribe Hominina, which includes all representatives of the human lineage, from extant modern humans to other extinct species of the genus *Homo* and of previous genera of extinct bipedal apes, such as *Australopithecus* and *Ardipithecus.*

Hominoidea The superfamily including apes, humans, and their direct and lateral extinct ancestors that are not also ancestors to the so-called Old World monkeys.

Hominoids Individuals belonging to the superfamily Hominoidea. There are currently only four kinds of hominoids in addition to the other apes, the Asian gibbons (*Hylobates*): *Pongo* (the orangutans of Borneo and Sumatra), *Gorilla*, *Pan* (chimpanzees and bonobos), and the last surviving species of *Homo.*

Homo The genus that includes extant humans, *H. sapiens,* and other extinct species such as *H. neanderthalensis, H. heidelbergensis, H. erectus, H. floresiensis,* etc. *Homo sapiens* is the only surviving species of this genus. The total number of extinct species is controversial.

Homo erectus The term is one of the first attributed to an extinct human species. Introduced at the end of the nineteenth century to classify forms such as *Pithecanthropus,* at the time considered to be half-ape (*pithēkos*) and half-man (*anthropos*), with upright posture and bipedal locomotion.

Homo heidelbergensis Middle Pleistocene hominins from Africa and Eurasia.

Homozygous mutation Identical mutation of both the maternal and paternal alleles.

Human Hominin belonging to the genus *Homo.*

Hyoid The only ossified element of the vocal tract, which can be preserved in the fossil record. It is held by ligaments and muscles that attach it to the mandible, temporal bone, thyroid cartilage, and sternum. It provides support to the larynx and an anchorage to the tongue and other muscles required for speaking.

Hypervariable region Part of the mitochondrial DNA that is characterized by high DNA sequence variability.

Interglacial A warm period between two glaciations.

Interstadial Brief warmer spell during a glacial period.

Isotope fractionation Effect produced by certain biogeochemical processes, such as evaporation, photosynthesis, and the precipitation of minerals, that select specific isotopes of an element, changing the original isotopic ratio (e.g., that of oxygen-18 to oxygen-16, hydrogen-2 to hydrogen-1, or carbon-13 to carbon-12).

Isotopes Forms of the same chemical element (same atomic number) with different atomic weights. They have different numbers of neutrons, but the same number of protons. Most chemical elements occur as mixtures of different isotopes. Radioactive isotopes are called radioisotopes.

Knuckle-walking The peculiar quadrupedal gait on the knuckles (upper limbs resting on the phalanges of four fingers) that characterizes the walking of current African apes.

Laser ablation inductively coupled mass spectrometry (LA-ICP-MS) Technique for elemental and isotopic microanalysis. The sample (e.g., the section of a tooth) is irradiated with a pulsed laser microbeam that drills holes of the size of a few hundred microns. This triggers mini-explosions that atomize the material, leaving tiny craters as the laser moves along the tooth's section. The atoms ejected from the sample are then channeled into a plasma flame that heats it to 7,000° C, ionizing it. Ions are then sent to the mass analyzer, which separates them according to their mass. This method can reach sensitivities at the part-per-billion level and provide high-resolution isotopic and elemental maps; but it is invasive, since the tooth must be sectioned.

Lineage A group of taxa with a common ancestor to the exclusion of other taxa.

Locus Location of a specific gene on the chromosome.

Lower Paleolithic The period between 2.5 MYA and 120 KYA (approximately).

Mean life Average lifetime of the nuclei of a particular unstable atomic species.

Megafauna Usually refers to large-bodied extinct vertebrates that existed in Australia and in other continents during the Pleistocene. Australian megafauna includes *Diprotodon optatum*, *Thylacoleo carnifex*, *Megalania prisca*, and *Genyornis newtoni*. The American megafauna includes *Camelops hesternus*, a mighty animal, more than two meters tall, that lived in North America since about 5 MYA.

Mesozoic The geologic era between 251 and 65.5 MYA.

Microblades Small stone blades made during the Upper Paleolithic from a wedge-shaped core. They were then retouched to create various forms of microliths.

Microlith Stone artifact obtained from the fragmentation of small splinters or blades. Sometimes microliths were obtained using silicates that had been heated to facilitate their trimming into specific shapes. These long and thin blades were probably placed in wooden rods to create innovative weapons for throwing.

Middle Awash Region that includes some ancient river valleys in the intermediate section of the course of the Awash river in Ethiopia.

Middle Paleolithic Period stretching between 300–120 and 40–35 KYA, depending on the geographic area in which artifacts of this type were found; lithic technologies of the early *H. sapiens* in Africa and in the Middle East also belong to this period.

Miocene The geologic epoch from 23.03 to 5.33 MYA.

Mitochondria Structures outside the nucleus that provide energy to the cell. They have their own separate genome, of circular form, which contains the instructions necessary for the various reactions that convert food into energy.

Mitochondrial DNA (mtDNA) Circular DNA in the mitochondria, which is often used for human evolution studies. Mitochondria are inherited only from the mother, without recombination, and accumulate mutations at high rates, providing an accurate molecular clock.

Molecular clock Method used to measure the time elapsed since two species shared a common ancestor. It is based on the assumption that genetic change happens at a constant rate over long periods of time.

Morphometric index The ratio between two measurements, such as (in anthropology) the so-called horizontal cephalic index, given by the proportion between width and length of the skull multiplied by 100. This simple calculation is at the basis of the famous distinction between dolichocephalics (individuals in whom the length of the skull tends to be almost twice its width, and therefore the skull is long and narrow) and brachycephalics (in whom the length and width of the skull approach each other so that the skull, seen from above, appears more rounded).

Mousterian Lithic industry based on scrapers and points that are not bifacial. It is generally associated with Neanderthals.

Multiregional evolution A theory of the 1940s, revisited in the early 1980s, supporting the idea that *H. sapiens* evolved from different groups of *H. erectus* (dispersed out of Africa 2 MYA) in several places throughout the world. This theory competes with the out-of-Africa hypothesis, which states that anatomically modern humans evolved only in Africa and dispersed from this continent about 70 KYA, replacing all archaic humans of ancient African origins.

Muon Elementary particle similar to the electron but with a mass that is 800 times larger.

Neolithic The last phase of the Stone Age, associated with polished stone tools, the beginning of agriculture, and the domestication of animals. In Western Asia agriculture started around 10 KYA.

Non-coding regions Regions of the mitochondrial DNA that do not carry genes but change rapidly and are also studied in population genetics.

Nucleus The central part of the cell, containing the chromosomes. Eukariotes (e.g., plants and animals) have nucleated cells, whereas prokaryotes (e.g., bacteria) do not. See also **Atom.**

Oldowan The first known lithic industry, associated with the appearance of *H. habilis.*

Oligocene The geologic epoch between 34 and 23 MYA.

Optically stimulated luminescence (OSL) Dating method based on the accumulation of energy in silicon and feldspar crystals found in geologic sediments, as a result of ionizing radiations from natural radioactive elements (uranium, thorium, and potassium) and secondary cosmic rays. The energy stored in the crystals is proportional to the time elapsed since the last exposure of the crystals to sunlight, which rapidly erases the luminescence signal and provides a starting time for the geological clock. In the OSL method, the electrons trapped in the crystal defects are made to return to their ground state by irradiating the crystal with a laser. The radiation emitted during this transition, which is proportional to the accumulated dose in the crystal, can be used to obtain ages of more than 500 KYA.

Orrorin The genus that includes late Miocene hominin remains found in Kenya.

Outgroup Monophyletic group of organisms used as a reference to determine the evolutionary relationships between different species.

Paleoanthropology The science that studies human origins and evolution.

Paleolithic The archaeological period that started 2.6 MYA in Africa with stone tools belonging to the Oldowan lithic industry; it ended around 10 KYA.

Paleomagnetism Dating method based on the changes of the magnetic field that took place in the past.

Phase contrast microCT MicroCT imaging based on X-ray phase contrast rather than conventional absorption. X-ray absorption gives low image contrast when the variations in density within the sample are small—for example, in the analysis of tooth enamel microstructure. In this case, it is advisable to use "phase contrast" imaging, which enhances the contrast by incorporating information on the X-rays phase. This method works very well with synchrotron radiation, which is characterized by a high "coherence" (X-ray waves have a fixed phase relationship between them). Very small changes in the phase of the waves crossing internal structures with small differences in density produce interference effects that improve image contrast.

Phenotype This term refers to the manifest character of an individual, a population, or a species, including behavioral aspects. It differs from the genotype, which deals more directly with genetic material.

Photons X-rays and other radiations can behave like a wave or a particle, according to the phenomenon observed; when they behave like a particle they are called "photons." The photoelectric effect, the emission of electrons by atoms bombarded with photons, was explained theoretically by Einstein; in 1921, he was awarded the Nobel Prize in Physics for this achievement.

Phytolith Deposition of silica in herbaceous plant cells that increases the rigidity of the stems and leaves. These materials are very hard, hence they are easily found in the fossil record. Their dimensions are variable, between 20 micron and 1 mm.

Pleistocene The geological epoch between 2,588 MYA and 11.7 KYA. The Lower Pleistocene is between 2,588 MYA and 781 KYA. The Middle Pleistocene is between 781 to 126 KYA. The Late Pleistocene is between 126 and 11.7 KYA.

Pliocene The geologic epoch between 5.3 and 2.6 MYA.

Polymerase chain reaction (PCR) A method to make copies of DNA molecules. First, the two DNA strands are separated from each other by heating the sample. Next, "primers"—small segments of nucleotides—are added along with the bases (A, C, G, and T) and a polymerase enzyme. When the mixture cools down, the primers bind to the ends of the DNA fragment that needs to be amplified, with each primer designed to adhere to only one of the paired DNA filaments. The temperature of the mixture is then raised and the enzyme starts to add bases to the primer, using the individual filaments of the sample as templates. The process is then repeated many times over, with the number of fragments of DNA between the binding sites of the two primers doubling during each cycle. A typical PCR run of thirty cycles can generate more than one billion identical DNA fragments (barring mismatches that introduce minor errors). Now present in large numbers, these fragments can be analyzed, for example by gel electrophoresis and subsequent visualization.

Positive selection The selection of a gene that is beneficial for selective purposes.

Protists Diverse group of eukaryotic microorganisms, which can be plant-like, fungi-like, and animal-like.

Quaternary The geologic period between 2,588 KYA and the present.

Racemization Structural transition between two versions of the same molecule.

Radioactivity The emission of ionizing radiation from an unstable nuclide (radionuclide). It includes beta particles (electrons), alpha particles (helium nuclei), and gamma radiation (such as X-rays, but with higher energies). Radioactivity was discovered in 1896 by French scientist Antoine Henri Becquerel, who noticed that uranium salts could expose a photographic plate. Marie Curie continued these studies with her husband, Pierre, using quantitative methods based on specific effects of radiation. Becquerel and the Curies won the Nobel Prize for Physics in 1903. In the following decades, using increasingly sophisticated methods and detectors, other natural radionuclides were discovered, some produced by cosmic radiation, others of primordial origin. Many have since been used as clocks in geology and archaeology.

Radiocarbon dating Dating method based on the decay of carbon-14 into nitrogen-14 (half-life of 5,730 years). Carbon-14 is produced in the atmosphere by cosmic-ray-generated neutrons and is taken up by living organisms. Its concentration in the atmosphere and living organisms is approximately one carbon-14 atom per one thousand million atoms of stable carbon isotopes (comprising carbon-12, 99 percent, and carbon-13, 1 percent). As the carbon-14 present in living tissues decays after death, the residual isotopic concentration of the remains provides us with the time of death. For example, after 50 thousand years from death, the carbon-14 concentration is about one thousand times lower than that of a living organism.

Recombination The creation of new genotypes from the regrouping of genetic material.

Retzius line Growth lines in tooth enamel that are produced during crown formation, named after Swedish anatomist Anders Retzius. Their external aspect on the dental surface is also referred to as "perikymata."

Ribonucleic acid (RNA) Involved in both protein synthesis and the transmission of genetic information. Unlike DNA, it usually comes single-stranded and in different shapes.

Sagittal crest A bone protuberance that runs across the skull superior midline for the attachment of large chewing muscles.

Sahelantropus tchadensis A hominin species hypothesized for a controversial fossil specimen discovered in Chad in 2001.

Sahul The supercontinent that existed during the ice ages, thanks to the low level of the oceans. It included the current continental Australia plus Tasmania and Papua New Guinea.

Savannah Tropical grassland with scattered trees.

Secondary ion mass spectrometry (SIMS) Method used to analyze stable isotopes. The mass of the isotope of interest is selected by transmitting the ions extracted from the specimen through magnetic and electric fields, which allows measurement of their velocity.

Secular radioactive equilibrium When a radionuclide decays into another radionuclide, after a period of time related to their respective half-lives, the production rate becomes equal to the decay rate, and the isotopic ratio between the two radionuclides remains constant. For example, as uranium-238 decays to thorium-234, the amount of thorium increases. After 500 thousand years, the rate of thorium decay becomes equal to the rate of its production, and the ratio between uranium-234 and thorium-230 remains constant. At this point the geologic clock does not work anymore.

Sexual dimorphism The difference in body size and morphology between a sexually mature male and a female of the same species.

Species A group of organisms that are classified together at the lowest level of the taxonomic hierarchy.

Spin One of the parameters that define the quantum properties of the electron. It is associated with its magnetic moment, which can interact with an external magnetic field. Two electrons, one with an "up" spin and one with a "down" spin, can occupy the same energy level. In the presence of a magnetic field, the energy level is split into two different levels, each occupied by one of the electrons with a different spin. Hence an unpaired electron is able to move between the two levels.

Sporopollenin Material composing the outer walls of pollen. Sporopollenin is a heteropolymer that resists chemical and enzymatic degradation, even with strong acids. It is composed of fatty acids, phenols, carotenoids, and phenylpropanoid.

Sunda The land including mainland Southeast Asia and the western Indonesian islands during the Pleistocene.

Supraorbital torus A bone bar extending over the superior margins of the orbits.

Synchrotron radiation High-intensity electromagnetic radiation produced by electrons moving near the speed of light in a doughnut-shaped ring at high vacuum, where their trajectory is bent by magnetic fields. The performance of synchrotron light accelerators is defined by a parameter called "brightness," namely the number of photons emitted per unit time, per unit area of the radiation source, per unit solid angle of the emitted radiation cone, and for a relative spectral bandwidth of 0.1 percent (photons/s/mm^2/mrad^2/0.1 percent BW). Since the discovery of synchrotron radiation in 1947, brightness has increased by several orders of magnitude. In a modern synchrotron, the brightness is more than 10^{22} photons/s/mm^2/mrad^2/0.1 percent BW (to be compared to about 10^7 photons/s/mm^2/mrad^2/0.1 percent BW emitted by conventional X-ray tubes). In "third-generation" synchrotrons the brightness is further enhanced with the inclusion of wigglers and ondulators, magnetic devices that force electrons along wavy trajectories.

Taphonomy The study of the processes that affect the state of the remains of organisms from their death time to fossilization.

Taurodontism Anomaly of the teeth characterized by an increased volume of the pulp chamber with associated reduction of the radical component.

Taxon A defined unit in the classification of organisms.

Taxonomy The rules and methods used to classify organisms.

Type specimen The specimen used to define the characteristics of a species, e.g., *H. neanderthalensis*.

Upper Paleolithic The period between about 40 and 10 KYA, in Europe, Northern Africa, and some regions of Asia. Archaeologists have identified during this period a sequence of different "lithic industries" or "cultures" of increasing complexity: the Châtelperronian, the Uluzzian, the Aurignacian, the Gravettian, the Solutrean, and the Magdalenian, often identified only in certain areas of Europe. The most ancient cultures, the Châtelperronian in France and Uluzzian in Italy, could represent periods of transition from the Mousterian of the Neanderthals.

Uranium series A dating method based on the radioactive decay of uranium. It can be used for direct dating of fossil teeth and bones.

Urey, Harold American nuclear chemist who won the Nobel Prize for Chemistry in 1934 for the discovery of deuterium, the heavy isotope of hydrogen.

Voxel Abbreviation for "volumetric pixel," the basic unit of the three-dimensional structure obtained in X-ray tomography.

X chromosome One of the sex chromosomes, present in one copy in men, and two in women.

Y chromosome One of the sex chromosomes, present only in males.

Years Before Present (BP) Conventional radiocarbon ages are reported in years before present (BP), where present is 1950 AD, and are calculated using the carbon-14 half-life of 5,568 years. They were originally used by W. Libby, the discoverer of the radiocarbon method, and are based on the assumption that the concentration of carbon-14/carbon in the past was constant; data are corrected for isotopic fractionation analyzing the ratio of carbon-13 to carbon-12 in the sample under observation. Calibrated (calendar) radiocarbon ages, which correct the conventional age to take into account the carbon-14/carbon variability in the past and use the correct radiocarbon half-life of 5,730 years, are reported as "cal BP."

Younger Dryas A short climatic event between 12.9 and 11.6 KYA, during the warming up that followed the Last Glacial Maximum. It takes its name from the *Dryas octopetala* plant growing in the tundra, which proliferated during that period to the detriment of the forests of Scandinavia.

Further Readings

Popular Books

Arsuaga, J.L. and I. Martínez. 2006. *The Chosen Species: The Long March of Human Evolution*. Malden, MA: Blackwell.

Cela-Conde, C.J. 2007. *Human Evolution: Trails from the Past*. Oxford, UK: Oxford University Press.

Finlayson, C. 2010. *The Humans Who Went Extinct: Why Neanderthals Died Out and We Survived*. Oxford, UK: Oxford University Press.

Lewin, R. 2005. *Human Evolution: An Illustrated Introduction*. Malden, MA: Blackwell.

Lieberman, D.E. 2011. *The Evolution of the Human Head*. Cambridge, MA: Harvard University Press.

Morwood, M.J. and P. van Oosterzee. 2007. *The Discovery of the Hobbit: The Scientific Breakthrough that Changed the Face of Human History*. Sidney, Aust.: Random House.

Stringer, C. 2011. *The Origin of Our Species*. London: Penguin Books.

Stringer, C. and P. Andrews. 2012. *The Complete World of Human Evolution*. London: Thames & Hudson.

Tattersall, I. 2012. *Masters of the Planet: The Search for Our Human Origins*. New York: Palgrave Macmillan.

Tuniz, C. 2012. *Radioactivity: A Very Short Introduction*. Oxford, UK: Oxford University Press.

Tuniz, C., R. Gillespie, and C. Jones. 2009. *The Bone Readers*. Sidney, Aust: Allen and Unwin.

Wood, B. 2005. *Human Evolution: A Very Short Introduction*. Oxford, UK: Oxford University Press.

Zimmer, C. 2007. *Smithsonian Intimate Guide to Human Origins*. Washington, DC: Smithsonian Books.

Selected Papers

Armitage, S.J. 2011. The southern route out of Africa: Evidence for an early expansion of modern humans into Arabia. *Science* 331, 453–56.

Austin, C., T.M. Smith, A. Bradman, R. Joannes-Boyau, D. Bishop, et al. 2013. Barium distributions in teeth reveal early-life dietary transitions in primates. *Nature* 498, 216–19.

Bastir M., A. Rosas, P. Gunz, A. Peña-Melian, G. Manzi, et al. 2011. Evolution of the base of the brain in highly encephalised human species. *Nature Communications* 2, 588. DOI: 10.1038/ncomms1593.

Berger, L.R. 2010. *Australopithecus sediba*: A new species of *Homo*-like Australopith from South Africa. *Science* 328, 195–204.

Briggs, A.W., J.M. Good, R.E. Green, J. Krause, T. Maricic, et al. 2009. Targeted retrieval and analysis of five Neanderthal mtDNA genomes. *Science* 325, 318–21.

Bronk Ramsey, C., R.A. Staff, C.L. Bryant, F. Brock, H. Kitagawa, et al. 2012. A complete terrestrial radiocarbon record for 11.2 to 52.8 kyr B.P. *Science* 338, 370–74.

Brown, P., T. Surikna, M.J. Morwood, R.P. Soejono, Jatmiko, et al. 2004. A new small-bodied hominin from the Late Pleistocene of Flores, Indonesia. *Nature* 431, 1055–101.

Bruner, E., G. Manzi, and J.L. Arsuaga. 2003. Encephalisation and allometric trajectories in the genus *Homo*: Evidence from the Neandertal and modern lineages. *Proceedings of the National Academy of Sciences USA* 100, 15335–40.

Brunet, M., A. Beauvilain, Y. Coppens, E. Heintz, A.H.E. Moutaye, and D. Pilbeam. 1995. The first Australopithecine 2,500 kilometres west of the Rift Valley (Chad). *Nature* 378, 273–75.

Cann, R.L., M. Stoneking, and A.C. Wilson. 1987. Mitochondrial DNA and human evolution. *Nature* 325, 31–36.

Caramelli, D., L. Milani, S. Vai, A. Modi, E. Pecchioli, et al. 2008. A 28,000 years old Cro-Magnon mtDNA sequence differs from all potentially contaminating modern sequences. *PLoS ONE* 3(7), e2700.

Cerling, T.E., E. Mbua, F.M. Kirera, F.K. Manthi, F.E. Groine, et al. 2011. Diet of *Paranthropus boisei* in the early Pleistocene of East Africa. *Proceedings of the National Academy of Sciences USA* 108, 9337–41.

Chappell, J. 2002. Sea level changes forced ice breakouts in the Last Glacial cycle: New results from coral terraces. *Quaternary Science Reviews* 21, 1229–40.

Condemi, S., A. Mounier, P. Giunti, M. Lari, D. Caramelli, and L. Longo. 2013. Possible interbreeding in late Italian Neanderthals? New data from the Mezzena jaw (Monti Lessini, Verona, Italy). *PLoS ONE* 8(3), e59781.

Dediu, D. and S.C. Levinson. 2013. On the antiquity of language: The reinterpretation of Neanderthal linguistic capacities and its consequences. *Frontiers in Psychology* 4, 1–17.

Di Vincenzo, F., S.E. Churcill, and G. Manzi. 2012. The Vindija Neanderthal scapular glenoid fossa: Comparative shape analysis suggests evo-devo changes among Neanderthals. *Journal of Human Evolution* 62, 274–85.

EPICA Community Members. 2004. Eight glacial cycles from an Antarctic ice core. *Nature* 429, 623–28.

Gillespie, R. 2008. Updating Martin's global extinction model. *Quaternary Science Review* 27, 2522–29.

Green, R.E., A-S. Malaspinas, J. Krause, A.W. Briggs, P.L.F. Johnson, et al. 2008. A complete Neanderthal mitochondrial genome sequence determined by high-throughput sequencing. *Cell* 134, 416–26.

Green, R.E., J. Krause, A.W. Briggs, T. Maricic, U. Stenzel, et al. 2010. A draft sequence of the Neandertal genome. *Science* 328, 710–22.

Hays, J.D., J. Imbrie, and N.J. Shackleton. 1976. Variations in the earth's orbit: Pacemaker of the ice ages. *Science* 194, 1121–32.

Henry, A.G., P.S. Ungar, B.H. Passey, M. Sponheimer, L. Rossow, et al. 2012. The diet of *Australopithecus sediba*. *Nature* 487, 90–93.

Henshilwood, C.S., F. d'Errico, R. Yates, Z. Jacobs, C. Tribolo, et al. 2002. Emergence of modern human behavior: Middle Stone Age engravings from South Africa. *Science* 295, 1278–80.

Henshilwood, C.S., F. d'Errico, M. Vanhaeren, K. van Niekerk, and Z. Jacobs. 2004. Middle Stone Age shell beads from South Africa. *Science* 304, 404.

Henshilwood, C.S., F. d'Errico, and I. Watts. 2009. Engraved ochre from the Middle Stone Age levels at Blombos Cave, South Africa. *Journal of Human Evolution* 57, 27–47.

Higham, T., T. Compton, C. Stringer, R. Jacobi, B. Shapiro, et al. 2011. The earliest evidence for anatomically modern humans in northwestern Europe. *Nature* 479, 521–24.

Hublin, J-J., S. Talamo, M. Julien, F. David, N. Connet, et al. 2013. Radiocarbon dates from the Grotte du Renne and Saint-Césaire support a Neanderthal origin for the Châtelperronian. *Proceedings of the National Academy of Sciences USA* 109, 18743–48.

Jacobs, Z., R.G. Roberts, R.F. Galbraith, H.J. Deacon, R. Grün, et al. 2008. Ages for the Middle Stone Age of Southern Africa: Implications for human behavior and dispersal. *Science* 322, 733–35.

Jouzel, J., V. Masson-Delmotte, O. Cattani, G. Dreyfus, S. Falourd, et al. 2007. Orbital and millennial Antarctic climate variability over the past 800,000 years. *Science* 317, 793–96.

Krause, J., Q. Fu, J.M. Good, B. Viola, M.V. Shunkov, et al. 2010. The complete mitochondrial DNA genome of an unknown hominin from southern Siberia. *Nature* 464, 894–97.

Lalueza-Fox, C., J. Krause, D. Caramelli, G. Catalano, L. Milani, et al. 2006. Mitochondrial DNA of an Iberian Neanderthal suggests a population affinity with other European Neanderthals. *Current Biology* 16, R629–30.

Lalueza-Fox, C., H. Römpler, D. Caramelli, C. Stäubert, G. Catalano, et al. 2007. A melanocortin 1 receptor allele suggests varying pigmentation among Neanderthals. *Science* 318, 1453–55.

Lalueza-Fox, C., E. Gigli, M. de la Rasilla, J. Fortea, A. Rosas, et al. 2008. Genetic characterization of the ABO blood group in Neandertals. *BMC Evolutionary Biology* 8, 342.

Lalueza-Fox, C., E. Gigli, M. de la Rasilla, J. Fortea, and A. Rosas. 2009. Bitter taste perception in Neanderthals through the analysis of the TAS2R38 gene. *Biology Letters* 5, 809–11.

Lari, M., E. Rizzi, L. Milani, G. Corti, C. Balsamo, et al. 2010. The microcephalin ancestral allele in a Neanderthal individual. *PLoS ONE* 5(5), e10648.

Lordkipanidze, D., M.S. Ponce de León, A. Margvelashvili, Y. Rak, G.P. Rightmire, et al. 2013. A complete skull from Dmanisi, Georgia, and the evolutionary biology of early *Homo*. *Science* 342, 326–31.

Manzi, G. 2012. On the trail of the genus *Homo* between archaic and derived morphologies. *Journal of Anthropological Sciences* 90, 1–18.

Manzi, G., F. Mallegni, and A. Ascenzi. 2001. A cranium for the earliest Europeans: Phylogenetic position of the hominid from Ceprano, Italy. *Proceedings of the National Academy of Sciences USA* 98, 10011–16.

Manzi, G., D. Magri, and M.R. Palombo. 2011. Early–Middle Pleistocene environmental changes and human evolution in the Italian peninsula. *Quaternary Science Reviews* 30, 1420–38.

Mellars, P.A. 2006. Why did modern human populations disperse from Africa ca. 60,000 years ago? A new model. *Proceedings of the National Academy of Sciences USA* 103, 9381–86.

Mellars, P.A. 2006. A new radiocarbon revolution and the dispersal of modern humans in Eurasia. *Nature* 439, 931–35.

Morwood, M.J., P. Brown, Jatmiko, T. Sutikna, E.W. Saptomo, et al. 2005. Further evidence for small-bodied hominins from the Late Pleistocene of Flores, Indonesia. *Nature* 437, 1012–17.

Noonan, J.P., G. Coop, S. Kudaravalli, D. Smith, J. Krause, et al. 2006. Sequencing and analysis of Neanderthal genomic DNA. *Science* 314, 1113–18.

Ovchinnikov, I.V., A. Götherström, G.P. Romanova, V.M. Kharitonov, K. Lidén, and W. Goodwin. 2000. Molecular analysis of Neanderthal DNA from the northern Caucasus. *Nature* 404, 490–93.

Peresani, M., M. Vanhaeren, E. Quaggiotto, A. Queffelec, and F. d'Errico. 2013. An ochered fossil marine shell from the Musterian of Fumane Cave, Italy. *PLoS ONE* 8(7), e68572.

Petraglia, M.D., A.M. Alsharekh, R. Crassard, N.A. Drake, H. Groucutt, et al. 2011. Middle Paleolithic occupation on a Marine Isotope Stage 5 lakeshore in the Nefud Desert, Saudi Arabia. *Quaternary Science Reviews* 30, 1555–59.

Petraglia, M.D., A. Alsharekh, P. Breeze, C. Clarkson, R. Crassard, et al. 2013. Hominin dispersal into the Nefud desert and Middle Palaeolithic settlement along the Jubah palaeolake, northern Arabia. *PLoS ONE* 7(11), e49840.

Raymoi, M.E. and P. Hybers. 2008. Unlocking the mysteries of the ice ages. *Nature* 451, 284–85.

Reich, D., R.E. Green, M. Kircher, J. Krause, N. Patterson, et al. 2010. Genetic history of an archaic hominin group from Denisova Cave in Siberia. *Nature* 468, 1053–60.

Reimer, P.J., M.G.L. Baillie, E. Bard, A. Bayliss, J.W. Beck, et al. 2009. IntCal09 and Marine09 radiocarbon age calibration curves, 0–50,000 years cal BP. *Radiocarbon* 51, 1111–50.

Roberts, R., M. Bird, J. Olley, R. Galbraith, E. Lawslett, et al. 1998. Optical and radiocarbon dating at Jinmium rock shelter in northern Australia. *Nature* 393, 358–62.

Roberts, R.G., T.F. Flannery, L.K. Ayliffe, H. Yoshida, J.M. Olley, et al. 2001. New ages for the last Australian megafauna: Continent-wide extinction about 46,000 years ago. *Science* 292, 1888–92.

Rosenberg, K. and W. Trevathan. 1996. Bipedalism and human birth: The obstetrical dilemma revisited. *Evolutionary Anthropology* 4, 161–68.

Rosenberg, T.M., F. Preusser, D. Fleitmann, A. Schwalb, K. Penkman, et al. 2011. Humid periods in southern Arabia: Windows of opportunity for modern human dispersal. *Geology* 39, 1115–18.

Simpson, S.W., J. Quade, N.E. Levin, R. Butler, G. Dupont-Nivet, et al. 2008. A female *Homo erectus* pelvis from Gona, Ethiopia. *Science* 322, 1089–92.

Stewart, J.R. and C.B. Stringer. 2012. Human evolution out of Africa: The role of refugia and climate change. *Science* 335, 1317–21.

Stringer, C.B. and P. Andrews. 1988. Genetic and fossil evidence for the origin of modern humans. *Science* 239, 1263–68.

Tafforeau, P., R. Boistel, E. Boller, A. Bravin, Y. Chairmanee, et al. 2006. Applications of x-ray synchrotron microtomography for non-destructive 3D studies of paleontological samples. *Applied Physics A* 83, 195–202.

Tuniz, C., F. Bernardini, I. Turk, L. Dimkaroski, L. Mancini, and D. Dreossi. 2012. Did Neanderthal play music? x-ray computed micro-tomography of the Divje Babe "flute." *Archaeometry* 54, 581–90.

Tuniz, C., P. Bayle, F. Bernardini, L. Bondioli, A. Coppa, et al. 2012. A new assessment of the Neanderthal child mandible from Molare, SW Italy, using x-ray microtomography. *American Journal of Physical Anthropology* 147 (Issue Supplement 54), 92.

Walter, R.C. 1994. Age of Lucy and the First Family: single-crystal 40Ar/39Ar dating of the Denen Dora and Lower Kada Hadar members of the Hadar Formation, Ethiopia. *Geology* 22, 6–10.

White, T.D., B. Asfaw, Y. Beyene, Y. Haile-Selassie, C. Owen Lovejoy, et al. 2009. *Ardipithecus ramidus* and the paleobiology of early hominids. *Science* 326, 75–86.

Yokoyama, Y., C. Falguères, F. Sémaha, T. Jacob, and R. Grün. 2008. Gamma-ray spectrometric dating of late *Homo erectus* skulls from Ngandong and Sambungmacan, Central Java, Indonesia. *Journal of Human Evolution* 55, 274–77.

Zollikofer, C.P.E., M.S. Ponce de León, D.E. Lieberman, F. Guy, D. Pilbeam, et al. 2005. Virtual cranial reconstruction of *Sahelanthropus tchadensis*. *Nature* 434, 755–59.

Index

About the Authors

Claudio Tuniz is an expert in advanced analytical methods using particle accelerators. He is scientific consultant (former assistant director) of ICTP/UNESCO in Italy and visiting professorial fellow at the Centre for Archaeological Science, University of Wollongong, Australia. Formerly, he was director of the Physics Division at the Australian Nuclear Science and Technology Organization. He has published extensively on applications of physical sciences in archaeology and paleoanthropology. He is coauthor of *The Bone Readers* (2010 *Choice* Outstanding Title) and editor-in-chief of *Archaeological and Anthropological Sciences*.

Giorgio Manzi is a paleoanthropologist acknowledged internationally for his studies on human evolution in Europe, with special reference to fossils such as those from Ceprano, Altamura, Saccopastore, and Monte Circeo. Well-known in Italy for his work as a science writer, he is professor of paleoanthropology, human ecology, and natural history of the primates at the University of Rome, La Sapienza, where he is also director of the Museum of Anthropology "Giuseppe Sergi." Since 2012 he serves as associate editor of the *American Journal of Physical Anthropology*.

David Caramelli is professor of anthropology and deputy director of the Department of Biology at the University of Florence. He also directs the Laboratory of Molecular Anthropology and Paleogenetics. He is an expert in ancient DNA analysis, and his research work is internationally recognized thanks to one hundred publications in the field of ancient DNA studies, which include several publications cited among the most important scientific reports for 2003 and 2007 (Science Breakthrough of the Year). He is associate editor of *BMC Genetics* and member of the editorial board of *PLoS ONE*.